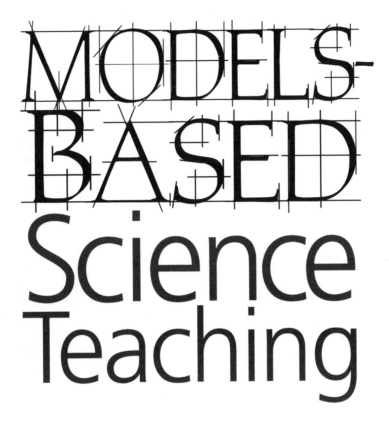

MODELS-BASED Science Teaching

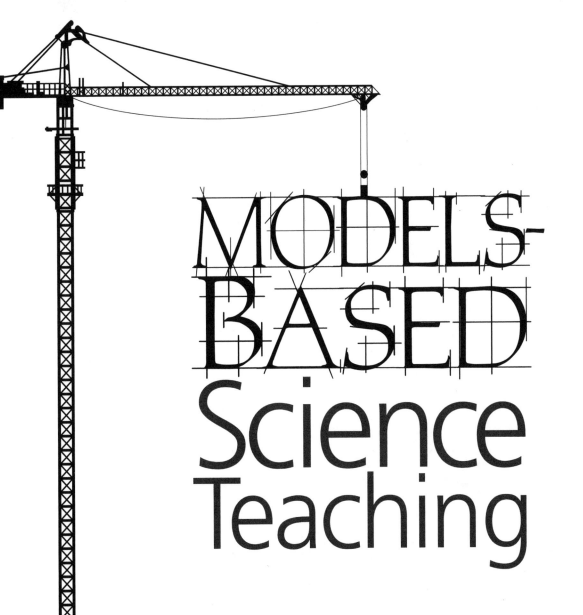

MODELS-BASED

Science Teaching

Steven W. Gilbert

National Science Teachers Association

Arlington, Virginia

National Science Teachers Association

Claire Reinburg, Director
Jennifer Horak, Managing Editor
Andrew Cooke, Senior Editor
Agnes Bannigan, Associate Editor
Wendy Rubin, Associate Editor
Amy America, Book Acquisitions Coordinator

ART AND DESIGN
Will Thomas Jr., Director

PRINTING AND PRODUCTION
Catherine Lorrain, Director

NATIONAL SCIENCE TEACHERS ASSOCIATION
Francis Q. Eberle, PhD, Executive Director
David Beacom, Publisher
1840 Wilson Blvd., Arlington, VA 22201
www.nsta.org/store
For customer service inquiries, please call 800-277-5300.

NSTA is committed to publishing material that promotes the best in inquiry-based science education. However, conditions of actual use may vary, and the safety procedures and practices described in this book are intended to serve only as a guide. Additional precautionary measures may be required. NSTA and the authors do not warrant or represent that the procedures and practices in this book meet any safety code or standard of federal, state, or local regulations. NSTA and the authors disclaim any liability for personal injury or damage to property arising out of or relating to the use of this book, including any of the recommendations, instructions, or materials contained therein.

PERMISSIONS

Library of Congress Cataloging-in-Publication Data
Gilbert, Stephen W.
  Models-based science teaching / by Steven Gilbert.
     p. cm.
  Includes bibliographical references and index.
  ISBN 978-1-936137-23-7
1. Science—Study and teaching—Methodology. 2. Concept mapping. I. Title.
Q181.G394 2011
507.1—dc23
                                2011029951

eISBN 973-1-936959-96-9

CONTENTS

CONTENTS

Chapter 6:

THE CREATIVE PROCESSES of SCIENCE

Chapter 7:

MBST and the SCIENTIFIC WORLDVIEW

PREFACE

Philosophy is dead.

—Hawking and Mlodinow, *The Grand Design*

The idea that philosophy is dead seems to be shared by many scientists and science teachers who believe that science has somehow supplanted all other approaches to the study of existence, knowledge, values, reason, mind, and language. In my view, however, nothing is further from the truth. Science is based on a commonly held philosophy, as are many other modes of formal human inquiry. Is philosophy dead in law? Economics? Politics? I think not. There are, in fact, many philosophies in the world today (probably as many as there are people, for we all have our personal philosophies). Unless we recognize and understand the strengths and weaknesses of those philosophies, we may be prone to make unnecessary errors in our interactions with the universe.

Models-Based Science Teaching (MBST) is not a primer on how to construct and use physical models for teaching. Instead, it builds upon the concept of mental models—simplified cognitive representations of what we think we know. From there, it seeks to demonstrate how we might lead our students to better define and frame the enterprise of science though models-based discourse, and thus how we might give more meaning to the processes and products of inquiry and discovery, research, and experimentation.

The key concept in this approach is mental modeling. Mental models are the essential elements—the building blocks—of intelligent communication, learning, knowledge, understanding, and action; and yet we pay only scant attention to them when we talk about science. This does our students a grave disservice. It deprives them of the deeper understanding of science they could acquire were we to incorporate mental modeling overtly into our approach to teaching and understanding science. This book explores how to do just that.

The philosophical concept underlying MBST—which I will call models-based science (MBS)—challenges conventional ideas about the nature of reality. The concept is not new or original, nor are its implications widely accepted by the general population, despite support for its premises from the medical, cognitive, and computer sciences. It would be understandable, then, for some science teachers to object to bucking their students' traditional ideas about the nature of reality in order to incorporate newer and potentially more controversial ideas into their science curriculum.

I was pleased, therefore, to find a model supporting MBS in Stephen Hawking and Leonard Mlodinow's recent book, *The Grand Design* (2010). In their work, they introduce a philosophical approach called "model dependent realism," or what I will refer to as MDR. MDR is very similar in it arguments and implications to MBS. You will not find references in this book to MDR by name, but you would recognize similar ideas if you read both books.

Both philosophical approaches focus on mental models and regard models as far more than convenient representational tools. Mental models are the heart and soul of conscious thought. All higher animals create and use them when thinking; humans, however, are consummate model builders, creating mental models with our minds and physical models with our hands, our voices, and our gestures. Through these models we create and represent our internal (subjective) realities. This is a simple idea, but it has profound implications—implications we must convey to our students if we are truly concerned with developing their science literacy.

Absent such learning, students may well grow up believing that reality is exactly what they perceive it to be: that their perceptions are real in an absolute sense. From that personal conviction, they may find it easy to believe that scientific knowledge is real in the same certain sense. And yet one of the central tenets of the nature of science is that nothing is absolutely knowable or provable. Why not? Well, that is what this book is about.

I have crafted this book for preprofessional students preparing to be specialized elementary/middle/high school teachers, elementary teachers of science, practicing science teachers at all levels who want to explore new and better ways to frame and model science, and parents or guardians who are homeschooling their children. Of course, anyone who educates or mentors new or practicing science teachers should also read this book.

Several short readings, suitable for secondary-level students, have been included in the appendix. While elementary- and middle-level students should learn MBS ideas through the discourse their teacher uses to guide inquiry and frame discussion, high school students may be introduced to the central tenets of this book in a more direct and thoughtful way, through these readings and subsequent discussions. Open discussion is always preferable to rote memorization, and students should be free to disagree or dispute some of the assertions in the readings.

In the end, the success of models-based science teaching will depend on your willingness as a professional science teacher to understand the underlying concepts of MBS/MDR and apply them as you direct, guide, discuss, lecture, or otherwise interact with your students. Consistency in application is most important if you expect students to understand and adopt a similar model as their own.

Reference

Hawking, S. and L. Mlodinow. 2010. *The Grand Design.* New York, NY: Bantam Books.

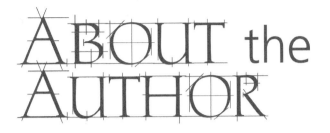

ABOUT the AUTHOR

Steven W. Gilbert received his PhD in science education from Purdue University in 1985 and has written extensively on models and modeling as a way to understand science. In addition to eight years of high school science teaching, he has been a professor of elementary and secondary science teacher education in Texas, Michigan, Indiana, and Virginia. He currently resides in Bloomington, Indiana, where he writes on science education and other topics. He can be reached at *stevengilb@gmail.com*.

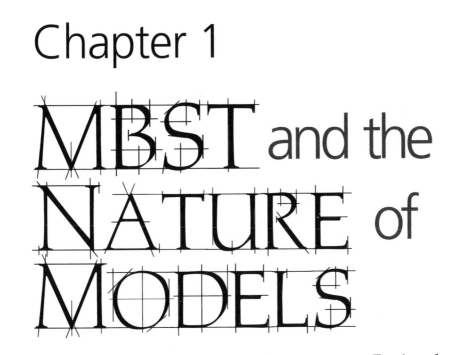

Chapter 1

MBST and the NATURE of MODELS

Equipped with his five senses, man explores the universe around him and calls the adventure Science.

—Edwin Powell Hubble

What Is This Book About?

This book is about modeling: specifically, modeling in science education. Although the Benchmarks for Science Literacy (AAAS 1993) recommended that teachers use models and modeling as a framework for building science literacy, this approach has only recently gained ground in the professional literature. A number of science educators have advocated the use of models as a framework for science education (Matthews 1988; Gilbert and Boulter 2000). Research-based projects such as "Modeling Instruction" (Jackson, Dukerich, and Hestenes 2008) claim considerable success in using models as a framework for conducting inquiry in high school physics. The movement toward explaining science through models and modeling is a logical outgrowth of developments in the cognitive sciences, where mental models are being used to conceptualize the processes and products of thinking.

In this book, we will explore the concept of mental models and how they relate to human learning—specifically, the learning of science. Then we will look at how we can apply this understanding when we teach science. Gobert and Buckley (2000) define models-based teaching as "any implementation that brings together information resources, learning activities, and instructional strategies intended to facilitate mental model-building both in individuals and among groups of learners." Building upon this definition, you can use models-based teaching to help your students develop a better understanding of what science is; how it is practiced and how it fits into the broader domain of their thinking. That is, how they can use model building for context.

By *context*, I mean the explanatory framework of science—it's gestalt. *Gestalt* refers to a unitary whole that is greater than the sum of its parts. The gestalt of science is a unifying conceptualization of science that goes beyond its subject matter content and its techniques of investigation alone to include the relationship of science to (a) the ways in which we think and know, (b) our beliefs about what is true, (c) our beliefs about science, and (d) our assumptions about the role of science in society and culture.

Gestalt combines process, content, and context to give science instruction optimal meaning. Your students will not acquire this gestalt by learning content knowledge and engaging in lab experiences. They also need an overriding framework to pull these activities together in a meaningful way.

Gestalt is important because of the links it establishes with other domains of human activity. Few of your students will practice professional science after they leave school. Some of your students will go their entire lives without meeting a practicing scientist. Your students will become journalists and writers; politicians and business leaders; workers, educators, administrators, environmentalists; lobbyists, bloggers, and, most important of all, voters. In these and other roles, their knowledge of the gestalt—the context of science—will be at least as important as their knowledge of specific science content and their classroom science experiences.

"But," you protest, "there are only so many hours in a school year! I teach them science for an hour a day, and in that time I'm expected to cover a lot of content. I do some inquiry, but inquiry takes time. How do I teach this gestalt of science without losing time for other things?"

That is one of the benefits of a good framework. How you contextualize normal science activities is at least as important as how you actually conduct them. Models-Based Science Teaching (MBST) requires only a small amount of additional instructional time. From that investment, you can reap big benefits.

MBST frames science as a formal process for constructing descriptive and explanatory models. It defines all learning as a process of modeling. This proposition, although simple, has profound implications for teaching the

- Process of scientific investigation

- Nature of the products of science (content)

- Purposes of science

- Limits of scientific knowledge

- Nature of learning and knowledge across fields

- Relationship between science and nonscience

Within this framework, we reap the added benefit of learning a great deal about ourselves as people. After all, we are all learners, whether or not we are scientists. Science is a process of learning—of modeling. So how is science different from the other ways that we learn? What claims can scientists really make about the knowledge they create? Can science really provide us with the ultimate answers to life, the universe, and everything?

To answer these questions, we first have to understand models. What are they? How do they work? Why do we use them? Once we have this background, we will be ready to proceed to a discussion of mental models, and their importance in human thinking.

Defining Models

What is a model? You may think you already know the answer to this question; your students probably think they know too, unless they are very young. But their ideas of what a model is, and perhaps your own as well, will be narrower than the definition I present in this book. Chances are, when I talk to you about models, that you first envision concrete objects meant to resemble other objects for which they stand, such as model airplanes, toys, globes, and plastic human torsos.

First off, we need to be on the same page about what a model is. The term *model* has several legitimate meanings. For our purpose, I will define a model as a system of objects, symbols, and relationships representing another system (called a *target*) in a different medium (modified from Gilbert and Ireton 2003). Models may be concrete or abstract, but they all share certain common properties; for example, they all function through analogy (Leatherdale 1974). Analogies are systems that have the general form, "A is to B as C is to D," shown in Figure 1.1. In this diagram, the features A and B (also called *attributes*) have the same relationship to each other as do the corresponding features, C and D, in the target. Attribute A corresponds to attribute C: while attribute B corresponds to attribute D. Otherwise, A/B and C/D are in different systems.

An example of a clearly analogical scientific model is Rutherford's "solar system" model of the atom, shown in Figure 1.2. The Rutherford model of the atom places the nucleus at the center of the atom, corresponding to the Sun at the center of the solar system. The electrons of the atom in orbit around the nucleus correspond to the planets orbiting the Sun (they have the same relationship). This is a simple model with an analogy that is easy to understand and therefore is effective, even if it's partially wrong.

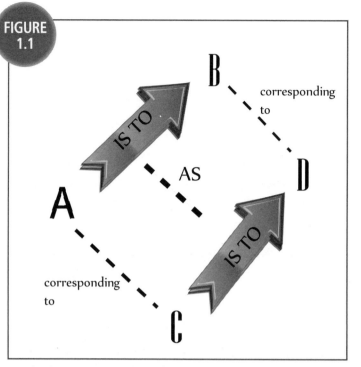

FIGURE 1.1

Analogies are comparisons between systems where the relationships between corresponding characteristics, or attributes, are the same or similar. All models are analogical to their targets.

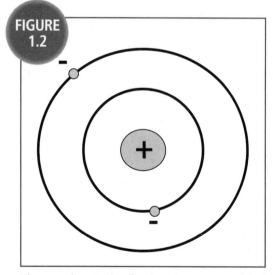

FIGURE 1.2

The early Rutherford Atom resembled our solar system, where the nucleus corresponded to the Sun, and the electrons correspond to the planets. Electrons and planets orbited the nucleus and Sun respectively.

The process of relating A to C and B to D is called *mapping*. For an analogy to work, we have to be able to map at least some attributes of the target onto the model and vice versa. If the mapping is easy for us, the model is highly *transparent*. The Rutherford atom model is transparent *if* you are familiar with the solar system.

Other models are not so transparent. For example, the cosmological model known as M-Theory (a.k.a., "String Theory") proposes that the universe is comprised of oscillating one-dimensional lines. What does a one-dimensional string look like? What's it made of? I don't really know. Neither, I suspect, does anyone else.

It's hard enough to visualize a one-dimensional line of some undefined composition, but M-theory also says that the universe has eleven dimensions (not the four we are familiar with in space-time). The additional seven dimensions are hard for the average person to understand because they have no familiar counterparts in the perceived world. They are difficult to explain in nonmathematical terms because they *are* mathematical models. The analogy to strings is simply a way to make the theory more concrete.

The string theory example clarifies the need for *rules of interpretation* through which we can define the relationships among the corresponding attributes of target/model systems. For scale models like model airplanes, the rules are usually simple. We recognize the visual similarity right away. But when models don't look anything like their targets, rules are needed. You must know, for example, the rules of interpretation for any mathematical model, even one as simple as "2 + 2 = 4." A "2" doesn't resemble anything in nature, nor does an "=" sign. These symbols represent abstract qualities (ideas) and nothing physical. You must know what they mean and how to use them.

Table 1.1 identifies several classes of common models. Some look nothing like their targets, but all of them are models under our previous definition. Like a landscape painting, each of them allows us to glimpse some isolated aspect of reality to which we might not otherwise have access; but none of them describes their target fully, nor should any of them be mistaken for their targets. This may seem intuitively obvious, but the confusion of models with their targets is a common source of misunderstanding.

The accuracy with which a model represents its target is called its *fit*. We construct a model for a particular purpose and its fit is a measure of how well we achieve that goal. A good model may be a poor fit if we use it for the wrong purpose. Fit and transparency are two different things. A highly transparent model may be a poor fit if it's used for the wrong purpose. The best road map in the world will probably not be a very good model for a pilot cruising at 10,000 feet.

All models are simplifications of their targets. Unnecessary details obscure the meaning of the model. But even more than that, a model is not necessary if you can work with the target. We make models because we cannot use or study the target usefully or practically in our situation. Most models are dramatically simplified, so that we are focused only on the elements that are most important to us.

Because models are simplified, they are always in some ways inaccurate and misleading. A road map, for example, cannot tell us much about the conditions of the road. A graph of an event is only marginally like the event in real time. Models always lack some information, and they often suffer from unavoidable

TABLE 1.1

Examples of Models in Several Categories

Class of Models	Examples
Concrete models	Scale models
	Mockups
	Figurines
Pictorial/graphic models	Blueprints
	Photographs
	Diagrams
Mathematical models	Formulae and equations
	Graphs
	Topographic maps
Verbal models	Descriptions
	Scripts
	Directions
Simulation models	Simulation games
	Cockpit simulators
	Crash test dummies
Symbolic models (semiotic models)	Words, numbers
	Mathematical figures
	Stoplights, stop signs

distortions caused by the change in medium. A typical road map does not tell you about hills, for example.

Despite these shortcomings, models are extremely valuable to us. In fact, as we shall see, they are indispensible. You may not be aware of it, but you construct models every day. These models exhibit all of the characteristics we have just discussed. They are purposeful, simplified, and often inaccurate and misleading. You are constructing one such model even as you read this page: it is a mental model through which you are making sense of these otherwise meaningless strings of symbols we call words.

Mental models are the foundations for understanding MBST. We will introduce and elaborate on them in the next chapter.

Summary

Models-Based Science Teaching is an approach to framing science built upon current metaphors and models in cognitive and computer sciences. In MBST, science is defined as a process of building descriptive and explanatory models of natural phenomena.[1] MBST incorporates into its own explanatory framework the theoretical notion of mental models.

Models are analogical systems that represent other systems, called targets. All models have certain defining characteristics: They are purposeful, simplified, and frequently can be misleading unless you understand the rules for interpreting them. Virtually anything we create to represent something else is a model.

Normally we strive to build models that are transparent and that fit the target best for our particular purpose. Any given target may be represented in different ways for different purposes. Mental models have the same characteristics that all models have.

For Discussion

1. Identify the target, parallel relationships, and corresponding attributes of (a) a world globe, (b) a street map, and (c) a line graph showing the change of water pressure with increasing depth.

2. Explain how each of these models is simplified and erroneous in comparison to their targets.

3. You are 5'6" tall and weigh 120 lbs. You are struck by an alien ray. Suddenly you triple in size in all three spatial dimensions. What happens to your weight? What problems might this cause you?

4. Examine the classes of models in Table 1.1. Chose one in each category as an example and explain how it meets the definition of a model we gave earlier. Map the primary corresponding attributes.

---

[1] A phenomenon is any observable event or occurrence.

References

American Association for the Advancement of Science (AAAS). 1993. *Benchmarks for science literacy.* New York, NY: Oxford University Press.

Gilbert, J. K., and C. Boulter, eds. 2000. *Developing models in science education.* Norwell, MA: Kluwer Academic Publishers.

Gilbert, S. W., and S. W. Ireton. 2003. *Understanding models in earth and space science.* Arlington, VA: NSTA Press.

Gobert, J. D., and B. C. Buckley. 2000. Introduction to models based teaching and learning in science education. *International Journal of Science Education* 22 (9): 891–894.

Jackson, J., L. Dukerich, and D. Hestenes. 2008. Modeling instruction: An effective model for science education. *Science Educator* 17 (1): 10–17.

Leatherdale, W. H. 1974. *The roles of analogy, model and metaphor in science.* Oxford, UK: North Holland Publishing Company.

Matthews, M. R. 1988. A role for history and philosophy in science teaching. *Educational Philosophy and Theory* 20: 67–81.

Chapter 2

MENTAL MODELS

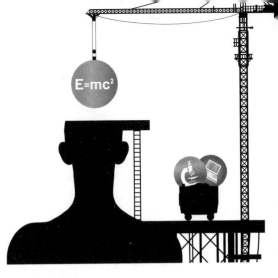

Science and mathematics/
Run parallel to reality, they
symbolize it, they squint at it,/
They never touch it …

—Robinson Jeffers

Mental models provide us with a framework for understanding MBST, as well as some of the tenets and limitations of science suggested by philosophers of science. But an understanding of mental models can take us beyond mere understanding of science. They are at the foundation of all learning, scientific or otherwise. So if we want to understand how science relates to other ways of knowing, we have to understand mental models.

This chapter is background reading for anyone who wants to use MBST. It will provide you with the models and the language you need to use MBST effectively in the classroom. I'm a great believer in the idea that teachers should understand why they are adopting a particular approach, as well as what that approach is. In this chapter and the next we will put learning, and then science, in a models context with the goal of understanding. In subsequent chapters, we will move our model into the science classroom.

Body and Mind

The question of how body and mind interact with each other has puzzled philosophers for centuries. To almost all Western thinkers, at least before modern times, the rational mind and the soul were one. To them, the soul had a real corporeal existence, often taking the form of the pale vaporous entity from which we derive our fictional concept of ghosts. To most classical philosophers, the only real question was whether the mind was separate from the body, as Plato (~400 BCE) argued, or was an inseparable part of the body, as Plato's pupil Aristotle (~350 BCE) taught.

The ancient Egyptians placed the seat of intelligence in the heart, but by around 450 BCE, Greeks such as Alemaeon argued that the mind resided in the brain. The highly influential Roman physician Galen (~150 CE), followed Aristotle's lead, teaching that the brain was the seat of the rational soul (mind) peculiar to humans; the heart was the seat of the animal soul (possessed by all animals and humans); and the liver was the seat of the vegetative soul (possessed by plants, animals, and humans).

Saint Augustine (~400 CE) accepted this notion of three souls (and with them the idea that plants and animals had lesser souls of their own). The rational soul—the mind—was of greatest interest to him. He argued that the mind combined the input from the five senses with divine illumination to arrive at truth. God revealed the truth of knowledge and knowledge had an objective existence beyond the body.

By the 16th century, the European medieval worldview was changing. For reasons we will discuss in the next chapter, a more mechanistic, less mystical, worldview evolved. The development of sophisticated machines led some intellectuals to a new metaphor explaining God: The watchmaker who set the world in motion and then stepped away from it. Men such as Andreas Vesalius (1514–1564), who dissected and studied the human body, began to question the models of Galen and other classical physicians. The body was looking more and more like a machine to them—one that they didn't understand, to be sure, but one without a comfortable resting place for the soul. Where did the three souls reside? Where was the rational mind?

The philosopher and mathematician Rene Descartes (1596–1650) was one of a number of seventeenth century scholars who struggled with the mind/body question. He accepted the notion that the body was a machine: a nonthinking thing. He also rejected the idea of three souls. To him, there was only one immaterial "thinking thing" that somehow interacted with the body but was separate from it, a model called *dualism*.

Descartes was unable to explain how an immaterial mind could interact with a material body. His views did not resolve the mind-body question, but they did make a convincing argument that the natural world—and our bodies with it—was a great machine. Unfortunately, his view of animals as machines without souls or the capacity for rational thought consigned these creatures to the twin hells of exploitation and vivisection. Dualism is still the basis for much western thinking, especially religious thought.

No one understood how the brain and nervous system worked at that time, nor was Descartes' idea of dualism acceptable to everyone. Early explanations for how our bodies worked often invoked mysterious forces that pulled levers and filled tubes to make our muscles and nerves work. Electricity was unknown as a source of power. Suggestions for how the brain functioned mirrored the knowledge of the day: Some suggested that the nerves were tubes conducting animal spirits, or fluids or air to control the body (an analogy with the circulatory system).

Rene Descartes (1596–1650)

The unresolved mystery of consciousness continued, and mystical explanations were widely accepted. And if the world was a machine, the human mind was still considered special and set apart from it. Machines, as understood in those centuries, offered no analogies to suggest how the brain produced consciousness. Only mechanical forces made sense to investigators at the time, and many organ functions remained a mystery. The natural philosopher Jan van Helmont (1579–1644) believed every organ had a soul, while the philosopher Thomas Hobbes (1588–1679) held that the heart was the center of the senses.

As science progressed into the 19th and 20th centuries, its explanatory metaphors changed. The function of the brain became better understood, but not how it worked. The mystery of mind-body dualism continued for most people. By the 20th century, a new metaphor arose that offered insight into the mind and consciousness: the metaphor of the computer. The computer provided us with a different view of a machine.

From nothing more than a series of electrically mediated ones and zeros, we could create imagery, store and process information, and create forms of artificial intelligence: in other words, we could create the *illusion* of reality. The brain, now much better understood, appeared to function in an analogous way, so much so that cognitive scientists today use computers to test their ideas about cognition (learning and knowing).

Of course, we have not resolved the mysteries of consciousness. We are far from knowing whether a computer can ever become conscious in the same sense as a human. But the metaphor of the computer does give us a rationale for placing our mental functions—our mind—in the brain and nowhere else.

Mind and Mental Models

What is this thing called *mind* anyway? In psychological parlance, the mind is the sum of all of our conscious and unconscious mental processes. These mental processes reside in the brain. Our rational thoughts appear to originate entirely within our brain. But though we are able to use instruments to chart the involvement of various parts of the brain as we think, the processes of thought—especially consciousness—remain mysterious.

Cognitive scientists generally view the brain as an organic computer that is preprogrammed to learn. The brain's input consists entirely of electrical signals carried to it though the nerves from our sensory organs. From these signals, our minds construct reality and then respond to that construction. The evidence establishing the body-mind connection is substantial. For example, if your brain suffers damage such that its programs no longer run normally, you may become mentally disabled. Drugs and alcohol affect your mental processes though explainable chemical interactions. We use antipsychotic drugs to treat certain mental diseases.

While it's possible to create an explanation for these observations that maintains the mind-body separation, such an explanation is necessarily more complex and tortured than the logical simplicity of the model seating the mind in the body. Few educated people blame gods or devils for mental illness anymore. The fact that brain damage can alter one's core personality implies that our personalities—the essence of who we are—are based in the brain and nowhere else.

Working from computer analogies, cognitive scientists initially hypothesized that the brain operated like a computer, executing complex commands using axioms, algorithms, and linear logical operations ("if A, then B"). Then, in *The Nature of Explanation*, Kenneth Craik (1943) suggested that the mind constructs small-scale mental models to anticipate and respond to future events. His notion of mental models languished, for the most part, until Johnson-Laird (1983) and Genter and Stevens (1983) published separate books with the same title: *Mental Models*. Since then, cognitive and computer scientists have used the concept of the mental model in their attempts to understand human thinking and design machines that that can better navigate complex environments.

The definition of a mental model varies in the professional literature. For our purposes in this book we will consider *mental models* to be distinguishable mental constructs that subjectively and holistically represent both objective external and abstracted internal reality.

Internal reality (i.e., *subjective* reality) is the perceived reality that we create within our minds, manifesting itself as the *streaming mental model* that we call consciousness, and as the *reference models* (memories) that our minds use as libraries.

The notion that we cannot truly know *objective reality*—the reality external to our senses—is not new. Plato and Pyrrho suggested the same thing in the fourth century BCE, but not for the same reasons that we suggest it today. Our current idea of mental models and the nature of reality is based upon what we know of perception and of the way the brain functions.

The Rationale for Mental Models

We receive all of our perceptions of objective reality through five senses—touch, taste, smell, sight, and hearing—and from internal sensors that inform our brain of our internal state of being. Our sensory receptor cells act as transducers to convert external energy into electrical signals that travel along the nerves to different areas of the brain to be processed (Figure. 2.1). Our brains then use this information to assemble the conscious image that is our streaming mental model, an analogy with streaming video. Of course, our streaming mental models—our conscious awareness—is only part of our total mental model. We are unaware of most of the work required to create subjective reality.

Picture yourself growing up in a spaceship. All your life, you have seen the outside world only through banks of electronic monitors (windows being too dangerous for use in deep space). The monitors look like windows and act like windows. You may not even be aware that they are monitors rather than windows, because those around you don't know that either. But an external observer, looking in, could see that your model of external reality is a created image. It is not a raw view of reality. Under normal circumstances, you don't need to know this. The monitors work. You can see the world around you. But you can never directly experience the world outside of your ship.

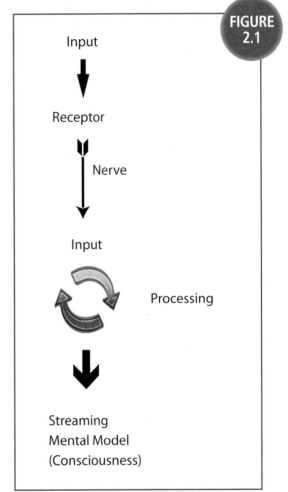

FIGURE 2.1

Input received by our sensory inputs goes to the brain as nervous impulses as raw information. Here it is translated and assembled into that of which we are conscious: our streaming mental model.

Our sensory experiences are based on incoming information, but our conscious models are created by our brain. For example, the property of color is not an objective property of electromagnetic energy. The electromagnetic spectrum is simply a continuous range of energy packets (quanta) that differ only in frequency and wavelength. These different wavelengths of light cause different signals to be sent to our brain from the color receptors in our eyes. The brain responds to these different impulses by imparting color to our streaming mental model. Not all animals see colors the way we do; in fact, we don't all see colors the same way.

Similarly, the odors we smell are not objective properties of matter; they are our subjective responses to chemicals. The chemicals themselves have no inherent property called "odor." Odors that repel us may be attractive to some animals (and to some other people). Sound is the subjective interpretation of waves of compression and rarefaction in the media around us. Without interpretation of our brain, sounds, odors, and colors do not exist.

If we didn't understand how the brain and nervous system operate, we could easily believe that our perception of external reality was unedited, requiring no translation. To the animal seeking merely to survive, an awareness of every thought process would be confusing and would slow down its mental processing without offering any corresponding benefit.

But for modern humans, especially for teachers and scientists, such awareness can be beneficial. Let's push our model a little further. Suppose you are striding down the hallway to your classroom after the bell has rung. You aren't feeling well and you fell asleep while taking a short break. The kids are in your classroom with only an aide. You see your principal walking toward you. She looks disturbed and upset upon seeing you. Both you and she know that you should be in your classroom. You feel guilty at once and you prepare an excuse.

Let's analyze this scenario. How do you recognize your boss? Why do you think she is in a bad mood? None of the raw information you are receiving from the external world answers these questions. Clearly, your streaming mental model, which includes all of these responses as well as a picture of the situation, is not just a raw image of what exists. As the model is

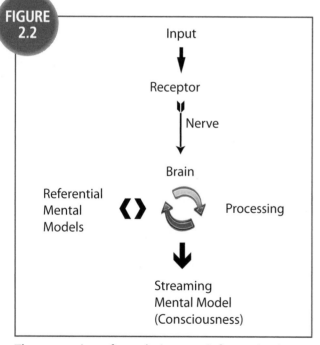

FIGURE 2.2

Input

Receptor

Nerve

Brain

Referential Mental Models

Processing

Streaming Mental Model (Consciousness)

The processing of translating raw information into a streaming mental model involves interactions with our referential mental models: our memories.

constructed, your mind uses your stored reference models to assess the current situation (Figure 2.2). You are not only informed, but you are prepared to act.

This little scenario illustrates that your subjective mental model contains much more information than you receive from your senses. In essence, your mind acts as a commentator and problem-solver, not just a reporter. It is more like a television writer/producer, making up the story and directing it. All streaming mental models are the products of subconscious interpretation. Interpretation, by definition, renders your reality subjective, not objective. To be conscious of a thing is to create and interpret an image of it. The very idea that you can know objective reality without subjective interpretation is unreasonable on its face.

If you are still not convinced that your reality is subjective, consider the research on how we receive information. At any given moment, you are aware of only a small portion of the information available from your environment. Even when you are focused on something, you do not receive information about it from your senses in a continuous stream. Instead, you receive information in bits and pieces and then reconstruct it (Bodovitz 2004). Your mind fills in the gaps by referring to its preexisting models, which themselves are incomplete by definition.

As you read this sentence, you do not perceive every letter in every word; nor do you consciously ponder the meaning of every word. If you look up and scan the room you are sitting in right now, you will see what I mean when I say you are merely sampling your environment.

The explanation for this sampling lies in the shortcomings of our nervous system. Unlike a television system composed of wires and mechanical transducers that do not need to recharge, nerve cells need to recharge because they get tired. Nerves are limited in the number of messages they can carry and the amount of information they can receive and process. By focusing on essential information informed by internal models, you can think faster. Such sampling causes errors, of course; therefore, we trade off speed for accuracy.

The internal nature of mental models is most obvious when we dream, when our streaming mental model evolves from the reference models in our mind. Dreams lack the discipline and purpose that our perceptions impose upon our waking models. This lack of discipline distorts them. They are usually elusive and vanish quickly from our minds once we awaken. In a purely physiological sense, though, they are as subjectively real as the streaming mental models you create when you are awake.

Remember that a mental model is not a concrete representation; rather, it is a representation of our perceptions. The best metaphor for how it operates is the computer. On most computers, the images and icons you see correspond in form to objects that are familiar to you, such as a page, file, or printer. These icons are nothing more than translations of signals originating in codes (ones and zeros) in the computer's memory where they have no recognizable form. In the brain, analogous codes reside in the nerve junctions of the brain (synapses).

The average human brain contains around one hundred billion nerve cells and a quadrillion nerve junctions (synapses) by age three (Drachman 2005)[1]. Changes in memory are associated with changes in these synapses, and translations of these codes give rise to our mental models. How this happens is still a mystery.

Incorporating New Information

Piaget (1954) was referring to a form of mental models when he talked about *schemata*. His definition of the *schema* (the singular form of schemata) was similar to, but more restricted in scope than, the mental model as we define it here. To Piaget, the schema was a mental representation of a physical or mental action (operation) that one could perform on an object, event, or phenomenon (Bhattacharya and Han 2001).

To understand Piaget's model more fully, consider what you have to do to navigate to your local coffee shop or to get to school in the morning. Your ability to navigate depends upon mental images representing significant places in your journey. You know when you have arrived at a familiar corner because the information that you receive from your senses maps onto your mental model of that corner. You don't call up this model consciously, in most cases, but you might become conscious of it if you were giving someone directions.

Your mind may associate that corner with an action or operation. Perhaps you must turn left there to get to the coffee shop, or right to get to school. These are not complex operations for us, and we might carry them out with minimal awareness. Our use of images in this way is routine. We don't think much about it. But the most modern computers would have a hard time replicating the navigation we do without effort.

What happens if we have to operate in a strange environment? Let's say we have to find a gas station in a strange town. We don't just search anywhere. Instead, we look for places where gas stations are located normally, such as at a busy intersection in a commercialized area. We know from experience that gas stations are found in certain places, and not in others. We can thus use this generalized principle to operate in towns that are unfamiliar to us. We will run into trouble only if the customs of a place are different from those to which we are accustomed. In that case, you must adapt by creating new images and rules. Mental models are constantly formed and modified as they are used. Recent studies show that the mere act of calling forth a memory can change it. Using Piaget's terminology, such *learning* requires either assimilation or accommodation. Let's look at what happens when you add new information to your existing mental model—when you learn.

If the new information fits into your current mental model, your mind will incorporate it by forming mental linkages (associations) with your existing reference models. This change requires relatively little mental energy or instability. Say, for example, you meet a new teacher in the school who will be working in another classroom. You try to remember her name, have a nice chat to be polite, and then go about your

---

[1]   The number declines with age.

business. You *assimilate* the information, but change none of your core beliefs, operations, or expectations. It's easy.

On the other hand, what happens if you are required to work with someone you dislike? This change creates a certain amount of mental turmoil. To work with this person, you have to alter your existing mental models. In this case, you have to *accommodate* the change. Accommodation requires more mental energy than assimilation and creates instability: an imbalance that Piaget called *disequilibrium*. Through accommodation, you try to reestablish equilibrium, or mental balance.

Biologists refer to equilibrium in the body as *homeostasis* (literally, "same state"). When we are comfortably at rest, we are in harmony with our environment. Environmental changes may force us to adjust until we achieve a new state of homeostasis different from the original state. How well we adapt to change in our mental model depends upon: (a) the nature of the change; (b) the alternatives available to us; (c) the costs/benefits of making a change; and (d) our conscious or subconscious assessment of the results of doing so.

Instead of accommodating change, though, we may resist or deny it. Denial is a well-known form of mental coping that can create temporary stability. If the change cannot be avoided, though, it will eventually require accommodation.

The Organization of Mental Models

Now that we have a reasonable model of mental models, let's discuss the way they are organized. Mental models vary in size, importance, and stability. The stability of a mental model depends upon how important it is to the overall mental structure, i.e., its centrality. *Centrality* is a product of the number and importance of the relationships (associations) that anchor onto it in the comprehensive network of models that makes up our worldview. We can call these central models *core mental models* (of CMMs, for short).

CMMs are, metaphorically, the most important supporting structures in the framework of ideas with which we operate. As any engineer or architect knows, you must be much more cautious about modifying a support wall in a structure than in modifying a nonsupport wall. The more the structure depends upon a given member, the more care you must take in changing that member.

CMMs are the essential determinants of who we are. They shape our goals, beliefs, explanations and decisions, and so they greatly influence our actions. For example, if our core mental model contains a strong component of trust, that trust component will spill over into many areas of our lives, as illustrated in Figure 2.3.

CMMs are born of many influences in our lives: the teachings of others, stories we hear, friendships we maintain, and events we hear of or witness. Because they support our total *worldview model*, we may resist modifying them. If we have to change them substantially, we may enter a period of mental instability—even crisis—before equilibrium is reestablished.

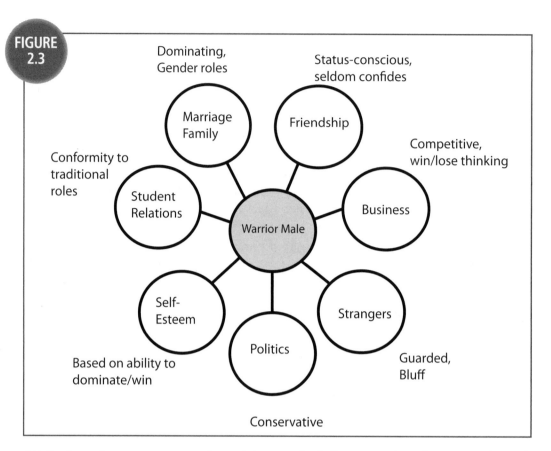

FIGURE 2.3

Dominating, Gender roles

Status-conscious, seldom confides

Conformity to traditional roles

Competitive, win/lose thinking

Marriage Family

Friendship

Student Relations

Warrior Male

Business

Self-Esteem

Strangers

Politics

Based on ability to dominate/win

Guarded, Bluff

Conservative

Attributes of our core mental models greatly influence other, less central models. A person with a strong tendency to trust will approach the world with different expectations, and make different decisions, than the person who is suspicious of everything.

Most core mental models form over time and are often rooted in childhood. Though usually stable, they may be altered significantly by brief but intense experiences such as fierce combat, violent crime, or some other event, giving rise to Post Traumatic Stress Disorder (PTSD).

Negative experiences are not the only things that can alter our CMMs. Positive experiences, such as a sudden and intense love relationship, can also alter our core models. Powers (2010) describes how a 13-year-old girl with whom a young Mark Twain spent three days as a young cub pilot influenced him and his writing throughout his life.

On the opposite end of the spectrum from the core mental models are what we might call *satellite mental models*. These are mental models that have little or no influence over our explanations and behaviors. Changing them requires few or no adjustment to our other models.

Satellite models are poorly integrated into our overall mental model. Unless you're a historian, for example, your mental model of gladiatorial combat in ancient Rome probably has little impact on your current worldview model. You could change it without much effort.

Our satellite models, though, are a kind of raw material. We hold on to them because they might be useful to us: for advancing our careers, for facilitating social interactions, or for solving problems. A lot of the models you learn in school are of this kind.

Why should we build mental models that we might never need? After all, learning requires energy and motivation. Our minds, consciously or subconsciously, must perceive value in learning something. While we might believe that we learn for the sheer joy of learning, there is usually more to it than that. We read for personal growth, but personal growth has an adaptive function. Reading a novel, for example, enlarges our experience, providing us with scripts that we might call upon to understand and make sense of our own lives. They give us something to talk about with friends (social function) or provide us with respite from reality (escape or recreation function).

Between our core and satellite models lies a range of mental models that are more or less integrated with one another. If we picture each mental model as a bubble, then each bubble represents a *concept*, which is, roughly, our idea of a thing. Concepts are interrelated by a complex web of associations.

We can illustrate these associations in various ways. One way is to represent each concept as a circle in a *Venn diagram*, such as the one in Figure 2.4. Each circle in a Venn diagram represents a conceptual model (images and relationships that comprise our idea of the thing), while the areas of overlap represent the percentage of elements that the models share.

Some conceptual models completely envelop others. In Figure 2.4, for example, the model of "mammal" completely envelops the models that represent "dog," "cat," and "human." Therefore, everything in our concepts of dogs, cats, and humans also applies to our concept of mammal. But the converse is not true: all of the characteristics of mammals apply to dogs, cats, and humans.

In this case, our models of dogs, cats, and humans *fuse* in one direction with the concept of mammal. Any changes we

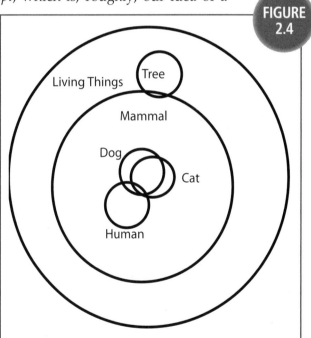

FIGURE 2.4

A Venn diagram works best for showing how some mental models are fused with others (encompassed by them) while other models are affiliated in a less complete way. Our mental models of mammal and tree are both fused with our model of living things, while our mental models of humans, dogs, and cats are affiliated.

make in the three lesser models change the greater model. I can think of no good examples of two-way fusion in models, since that would mean that the models were the same thing.

Fused models are on one end of a *spectrum of affiliation*. On the other end are poorly associated models.

Conceptual mental models form a natural hierarchy. We associate the concept of dog more closely with the concept of cat than with human. The concept of "tree" is out in left field, only marginally associated with the mammal.

One way to illustrate hierarchy is with a *concept map* like the one shown in Figure 2.5. In a concept map, we can see the hierarchy better, but we lose the information about degree of overlap.

Each model has its purpose and its trade-offs. Regardless of which one we use, they both simplify the organization of mental models in our minds.

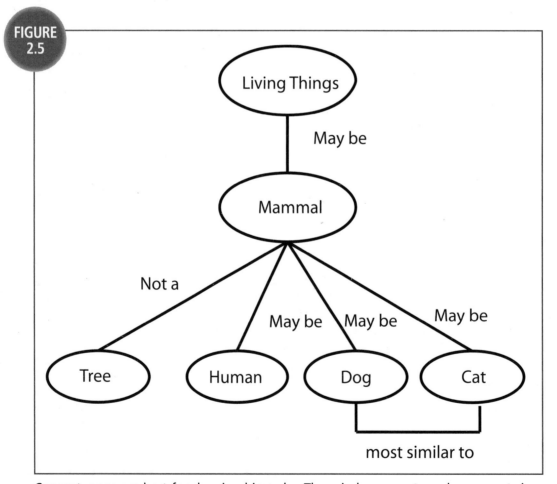

FIGURE 2.5

Concept maps are best for showing hierarchy. The mind appears to order concepts in a hierarchical manner through the associations it constructs.

Mental Models and Learning

Imagine yourself at the beginning of your conscious life, swimming in a sea of undifferentiated perceptions. Gradually, the object and events around you acquire distinctive forms, sounds, shapes, smells, tastes, and textures. Without any outside help, your mind begins to sort and categorize them. Certain experiences become associated with pleasure and happiness, while others become associated with fear and discomfort. Still others have no meaning for you at all.

From the moment of birth (and perhaps before then), we differentiate visual, tactile, auditory, gustatory, and olfactory mental representations—*sensory images*—in our minds. Along with these images come associations. These associations differentiate the concepts we form. They tell us what to include in each of our mental models and what to exclude. Each concept is a category into which we sort our experiences: all of the objects that we call dogs, for instance.

As we learn and grow, we acquire signs and symbols to represent or label elements of our shared models. Most animals communicate through signs and symbols with simple associations, by howling, scent marking, grooming, and so forth. Animals such as dogs, which engage in relatively complex behaviors, seem capable of reasoning from relatively sophisticated images, signs, and associations, as do young children.

Humans, of course, have evolved mental capacities beyond those of dogs and other mammals. Somewhere along the line, we evolved the capacity to organize symbols such as words, numbers, and signs into a syntactical structure—like those found in this sentence. *Syntax* refers to the arrangement of symbols into an order that gives the whole system meaning beyond the individual meanings of each symbol separately. Birds and whales exhibit what is called song syntax, but this is a meaningful ordering of otherwise meaningless sounds. It is not to be confused with linguistic syntax, in which the individual sounds (words) also have meaning.

Humans appear to be nearly alone in their ability to communicate linguistically. Quattara et al. (2009) have reported evidence of the use of proto-syntax in Campbell's monkeys. Other animals such as dogs, cats, monkeys, and parrots understand certain sounds in our speech and may seem to understand whole sentences, but the jury is out on whether they can understand syntactical communications in that way.

Parrots, for example, can mimic human speech and can associate specific words with specific things, but the associations are limited. It's doubtful that most other animals process meaning syntactically.

Humans have extensive ability to create original syntactical linguistic and mathematical models. Because these models are expressing mental models, we can group them together as *expressed models*. Any model in any medium that is constructed to express a mental model falls into this category.

In both language and mathematics, we organize symbols into syntactic *propositions* expressing relationships that may be true or false, and that exist no matter how we express them. For example, the proposition "force equals mass times acceleration," means the same thing as *F=ma*. It means the same thing even if it is expressed in Russian. The proposition is the underlying meaning of the phrase, not its expression (King, 2001).

Put another way, written or spoken propositions are declarative associations among the concepts they represent. In syntactical communication, the order of the symbols is important. *F=ma* is not the same thing as *a=mF*. "John does no harm" does not mean the same thing as "No John does harm," even though the symbols (words, in this case) are the same.

In addition to images, propositions are major components of human mental models. We may remember some of them as verbal images and create others as we need them. However, they must always be created internally before we can express them externally.

Propositions and Parsimony

Genter and Stevens (1983) point out that our mental models tend to be parsimonious, meaning that they are generally stored in the simplest form possible. This doesn't mean that our mental models are simple, of course; merely that they are simplified.

A good example of mental parsimony is found in our innate tendency to fear snakes and spiders. Most people who happen upon a snake will freeze before they consciously react, most often with innate caution. According to Öhman and Mineka (2001), "the evidence … shows that snake stimuli are strongly and widely associated with fear in humans and other primates and that fear of snakes is relatively independent of conscious cognition." They propose that most primates are born with an innate *fear module* regarding snakes. Humans may also naturally fear spiders and similar creatures that seem alien and scary to us.

These modules are not mental models of a specific snake or spider, but, rather, are of a general archetype of snakes and spiders that can lead us, for example, to jump away from a stick that looks somewhat like a snake before we can determine that it is not a snake. Fear modules can be modified through experience, so they differ from reflex. Such a module is an example of natural parsimony in our mental model structure.

We would be overwhelmed if we kept all of the information that we receive from the outside world. So when we receive worthwhile new information, our minds pare it down and sculpt it to fit our existing mental framework.

One way we simplify is to create categories of similar events. If the events are essentially equivalent, we can treat them in the same way, eliminating a good deal of unnecessary mental processing. How this is done is not clear, but it seems likely that the brain creates general pattern models similar to the fear module that we just discussed and associates it with certain responses.

These general models usually contain *generalizations and principles*, the latter of which are our personally accepted truths. They form the skeleton of ideas supporting our worldview model. Some of these principles are known to us. Others may be hidden. We may, for example, state a guiding principle such as "honesty is the best policy," while at the same time acting on a hidden belief that "omission of facts is not a lie."

Some principles are so central to our overall mental model that we simply assume that they are true. These principles are called, appropriately enough, *assumptions*. We may or we may not be aware of them when we act, and if we act on the wrong assumption, the results can be disastrous. Nevertheless, parsimony demands that we make assumptions.

Rules are mental directives that follow from certain principles and assumptions. A rule is just what the term implies: a prescribed response to a particular situation. We often follow rules with little thought—which, again, is a reflection of mental parsimony.

Scripts and scenarios are more holistic elements of our mental models. A *scenario* is an outline, while a *script* is a more detailed plan for action. We've all created scenarios and scripts by mulling over what we should say or do in a particular situation. Such holistic planning is—again—a reflection of mental parsimony, since scripts allow us to act rapidly and efficiently, without a lot of intervening thought.

Although we may create scenarios and scripts deliberately, as we do when we plan a speech or presentation, many scenarios and scripts evolve as a response to experience. When we meet people for the first time, for example, we tend to follow a script as we introduce ourselves. Experienced teachers follow previously effective scenarios when they interact with students.

Mental models contain many other elements as well: goals, objectives, values, ethics, exemplars, expectations, and so on. The main point for now is that these elements are often simplified. We forsake most of the details of experience over time, except for those we may retain as examples. As we change our models—which we do all the time—older elements may become nebulous, fragmentary, and confused. Such confusion is normal, but it may lead two people to describe the same event in much different ways.

Despite the problems it can cause, parsimony is adaptive in purpose, allowing us to act quickly when quick action is called for. But trimming the details also increases the risk of making a bad decision in an unfamiliar situation.

Truths, Archetypes, and Creativity

"You know," Arthur said thoughtfully, "all this explains a lot of things. All through my life I've had this strange unaccountable feeling that something was going on in the world, something big, even sinister, and no one would tell me what it was."

"No," said the old man, "that's just perfectly normal paranoia. Everyone in the universe has that."

—Douglas Adams, *The Hitchhiker's Guide to the Galaxy*

Some readers may resist accepting the notion that our minds are as mechanical as I have portrayed them. Like Mark Twain, who lamented the loss of the romance of the river when he learned how to be a riverboat pilot, you may mourn the loss of mystery and romance of more mystical explanations.

But our modern concept of mental models still does not answer the primal mystery of how consciousness and awareness has arisen from nonliving matter. The ultimate nature of reality, whether objective or subjective, lies beyond modern scientific explanation. Even if we learn the mechanics of how consciousness operates, that does not answer the question of how and why it evolved.

We recognize speculative ideas when we organize our ideas according to their "truth value." Saint Augustine believed that God provides divine illumination that allows us to recognize Truth with a capital T—an ultimate Truth separate from our physical existence.

Most cognitive scientists and realist philosophers would use a different definition of truth, preferring one with a small t. In their model, *truth* refers to the alignment of what we think we know (our mental models) with the facts that we perceive in the objective world.

The key issue in this case is this: since our mental models are subjectively constructed, how can we ever know what is true? Can science lead us to absolute Truth with a capital T? The answer is, probably not.

As I write this book for you, I can only assume that you and I share the same objective reality. I assume that because, if I hand you a pencil, you take it. The pencil is real to me, and I assume, to you. I can infer that my dog shares my objective reality for the same reason, though I have no idea what she is thinking. If I give her a treat, she takes it. We are communicating.

I know the world in three ways. First of all, I maintain a set of models that to me represent objective reality. I refer to these when I begin a sentence, "Well, realistically…." These I can refer to as my set of *reality models*. To you, my reality models may be wrong, or untrue.

The goal of science is to create reality models upon which we can all agree.

I also have in my worldview a set of models representing what I think ought to be. I can think of these as my *ideal models.* These are my standards, and they represent the world as I think it should be. I use these models to evaluate what I perceive happening.

My third set of mental models I know to be fantasies; I can call these my *fantasy mental models.* I create them through a process called *conceptual blending* (Fauconnier and Turner 2002).

For example, most of us have fantasized about achieving wealth and fame. In our daydreams, we imagine ourselves in a different frame of reality: we blend our self-concept with the characteristics we attribute to wealth and fame to create an imaginary model. Humans are probably unique in their capacity for creating ideal and fantasy models.

These categories of mental models often overlap, leading us to confuse fantasy models with realistic ones and to pursue unworkable and unattainable ideal models. Our models of the past may thus morph into nostalgic fantasies, and our plans for the future may founder upon reefs of improbable expectations shaped by our ideals.

Models of all three types are often shared across cultures. The mere fact that a culture shares a similar model does not make that model objectively true (aligned with facts), and hence real. After all, if I grow up believing in a fantasy model, and everyone else around me shares that belief, how will I know it is a fantasy? I will only know when that fantasy fails me in some way. Then I have to look for another model to replace it.

It seems probable, based upon our differences in experiences, that no two people hold exactly the same mental model of anything. Even the most casual conversations usually reveal many spoken variations in the same mental models of the conversationalists—and there are probably far more unspoken variations.

With so much variation, it is a wonder that we humans can agree on anything. And if this seems to you so far like a lot of philosophizing, let me point out its significance for teaching science (or anything else, for that matter):

- All that we think we know is only a mental model of what might actually be.

- The certainty of our models only comes from our determination of how well they work.

- Science is our most productive way yet of building reality models.

- Scientific or not, no one understands a model of any complexity in exactly the same way.

Although you may have been brought up to believe that we humans can know the certain Truth of things, indications are that such a belief is deceptive: a fantasy. At best, we feel our way through life by sampling, estimating, simplifying, and assuming; and then by creating a story.

Intellectually, we will benefit if we understand and separate ourselves from our stories so that we can examine them and assess their truth-value. If we know our beliefs are models, we can look at those models realistically: as simplified constructs that we have put together and that we can reject if they make no sense to us, or if they do not align with the truth we perceive in the facts.

Likewise, if we want to understand science, we cannot think of it as a pathway to unshakable Truth. Modern notions of the nature of scientific knowledge as found in the literature—scientific knowledge is changeable [tentative], empirically based, subjectively derived, socially embedded, and constructed (as opposed to discovered) (Clough 2007)—are consistent with theoretical notions of mental modeling. We will develop the rationale for this contention in the next and subsequent chapters.

For Discussion

1. Not everyone would agree with this model of how we think. What other models of thinking are held by people you know? Can we gain knowledge in any ways other than through our five senses?

2. Clarify and then discuss any concerns you have about the concepts of subjective and objective realities as defined here. Do you accept that objective reality is essentially unknowable? If not, why not?

3. Discuss the following statement: "Knowledge does not exist except in the individual's mind. If you remove the individual mind, all you have left is a bunch of sounds, dots, and squiggles." What do we mean by this? Do you agree or disagree? Why?

4. What is meant by the following statement: "We can share information, but not knowledge." Define information and knowledge as used in this statement. Do you then agree or disagree? Why or why not?

References

Bhattacharya, K., and S. Han. 2001. Piaget and cognitive development. In *Emerging perspectives on learning, teaching, and technology*, M. Orey, ed. Athens, GA: University of Georgia. *http//projects.coe.uga.edu/epltt/*

Bodovitz, S. 2004. Consciousness is discontinuous: The perception of continuity requires conscious vectors and needs to be balanced with creativity. *Medical Hypotheses* 62: (6), 1003-1005.

Clough, M. P. 2007. Teaching the nature of science to secondary and post-secondary students: Questions rather than tenets. *Pantaneto Forum* 25. *www.pantaneto.co.uk/issue25/clough.htm*

Craik, K. J. W. 1943. *The nature of explanation.* New York, NY: Cambridge University Press.

Drachman, D. 2005. Do we have brain to spare? *Neurology,* 64 (12): 2004–2005.

Fauconnier, G., and M. Turner. 2002. *The way we think.* New York: Basic Books.

Genter, D., and A. L. Stevens, eds. 1983. *Mental models.* Hillsdale, NJ: Lawrence Erlbaum Associates, Inc.

Johnson-Laird, P. N. 1983. *Mental models.* Cambridge, MA: Harvard University Press.

King, J. C. 2001. Structured propositions. *Stanford encyclopedia of philosophy. http://plato.stanford. edu/entries/propositions-structured*

Öhman, A., and S. Mineka. 2001. The malicious serpent: Snakes as a prototypical stimulus for an evolved module of fear. *Association for psychological science. www.psychologicalscience.org/ journals/cd/12_1/Ohman.cfm*

Piaget, J. 1954. *The construction of reality in the child.* New York: Basic Books.

Powers, R. 2010. Mark Twain in love. *Smithsonian* (May): 78–84.

Quattara, K., A. Lemasson, and K. Zuberbüler. 2009. Campbell's monkeys concatenate vocalizations into context-specific call sequences. *Proceedings of the national academy of sciences. www.pnas.org/content/106/51/22026.full*

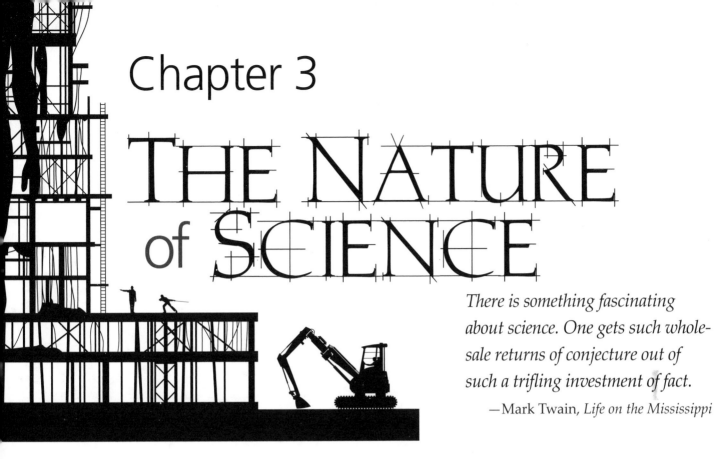

Chapter 3

THE NATURE of SCIENCE

There is something fascinating about science. One gets such wholesale returns of conjecture out of such a trifling investment of fact.

—Mark Twain, *Life on the Mississippi*

ark Twain's wit notwithstanding, science has had an enormous and largely positive impact on modern society, providing technologists with explanations leading to new inventions and changing our ideas about our world and our place in it.

While most people appreciate the contributions of science to our quality of life, some also worry about it. Science seems to them like a coldly rational enterprise that robs life of its mystery and romance. Not only that, but it poses a threat to the environment and to traditional societal values.

Thomas Huxley (1901) noted that the great tragedy of science was the slaying of a beautiful hypothesis by an ugly fact. True enough. Certainly, science has caused us to question and reject some beautiful fantasies; but it has allowed us also to slay fearsome dragons.

This chapter will provide you, as a teacher, with an overview of science: where it comes from, what it is, and how it relates to the MBST model. Obviously, a single chapter will not allow us to address everyone's questions and concerns. We will necessarily focus on the big picture. And because scholars disagree on elements of the theoretical nature of science, you may find others who disagree with some of my model. Such is the nature of the models we build.

We will begin this chapter by defining science and exploring its relationship with its first cousin, technology. The two fields are often confused. I will lay out reasons for keeping them separate.

We will go on to discuss the history and development of science to provide a context for science in modern society. Finally, I will present a brief model of the practices of science today, which I will enrich in future chapters.

Defining Science and Technology

The modern relationship between science and technology leads many people—including professionals in both fields—to treat them as though they were the same thing. In truth, the distinction between them is not as clear as the National Science Education Standards (NRC 1996) asserts. According to the NSES, scientists seek to describe and explain phenomena, while technologists such as engineers seek to invent and to apply knowledge to solving practical problems.

Consistent with the NSES and our MBST framework, we will define *science* as a system of tenets and methods for constructing and verifying descriptive and explanatory models of natural phenomena. More specifically, scientists build mental and expressed models to enrich their theoretical mental models. Theoretical models (a.k.a. *theories*) are stable and well-supported mental models that explain phenomena.

Technology, in contrast, is a system for inventing technical means (techniques) to achieve an end, frequently but not exclusively with an eye toward solving practical problems. As we will discuss later, though, there are theorists who regard science as a form of technology. I will draw a distinction between the two, as do the national standards.

History teaches us that inventors and engineers do not necessarily need to understand the explanation (theory) behind a phenomenon in order to develop a technique and make use of it. The medieval Chinese were able to compound gunpowder without understanding theoretical chemistry and early metallurgists discovered and manufactured bronze and iron without understanding the atomic structure of metals and alloys. Mathematicians charted the heavens and motions of the stars and planets without understanding the causes of their motions and physicians developed techniques for treating illnesses and wounds without understanding how the body worked.

It's worth examining the evolution of human thinking over the millennia in order to better understand, and perhaps better appreciate, the special but different places that technology and science have occupied in human cultural development. No one knows why modern technology developed so suddenly after such a long lag phase. Human technological activity in the form of stone tool making dates back around two million years (Plummer et al. 2009). The first known evidence of permanent stone structures appeared around 12,000 years ago in what is today Syria, when people planted crops, domesticated animals, and entered into a new phase of human development (the Neolithic or New Stone Age),[1] distinguishable by artifacts of a more settled existence, such as pottery.

Based on what we know about modern groups who live a similar existence, we can assume that cycles of nature, tradition, myth, and magic governed the worldviews of these people. Shamans, witch doctors, and healers practiced magic, propitiated the gods, and healed the injured and sick. Warriors hunted, herded, raided, and defended territory; and women wove, farmed, gathered foods, and cared for children.

Models that explained existence, both practical and sacred, were passed down orally through the generations. Pictures and symbols were used to tell stories, bring good fortune, and ward off evil. People believed that only a thin veil of illusion separated them from capricious gods and spirits—especially spirits of the dead—that lived all around them. The only way they could avoid trouble was to placate them with ritual and sacrifice. Fate governed life, and humans were largely helpless against it.

Between 8,000 and 5,000 years ago, we find the earliest signs of state-level societies (*civilization*) in Mesopotamia, which is modern Iraq (Kennett and Kennett 2006). The Neolithic period ended in the Middle East around 5,300 years ago when bronze, an alloy hard enough for weapons and tools, was discovered. Although anthropologists have found counting systems dating back 37,000 years, systems of mathematics and writing came into use only around 5,500 years ago (Melville 2003).

In the great scheme of human evolution, civilization exploded onto the scene in an instant, largely on the basis of technological advances. Drawings and pictographs evolved into systems of writing and mathematics. Healers, shamans, and witch doctors became philosophers, priests, and physicians. Castes grew up. Certain technologies flourished, but supernatural (rather than scientific) explanations persisted:

> The practice of magic was omnipresent in classical antiquity. The contemporaries of Plato and Socrates placed voodoo dolls on graves and thresholds… Cicero smiled upon a colleague who said he had lost his memory under the influence of a spell, and the Elder Pliny declared that everybody was afraid to fall victim to binding spells. The citizens of classical Teos cursed with spells whoever attacked the city; the twelve tables legislated against magical transfers of crops from one field to another; and the Imperial law books contain[ed] extensive sanctions against all sorts of magical procedures—with the sole exceptions of love spells and weather magic…. Magical rites gave access to a higher spirituality. These rites could open the way to the supreme god. (Graf 1999)

---

[1] We've skipped the Mesolithic for convenience.

In a world that seemed at the mercy of capricious and unstable spirits, the primary purpose of learning was to gain control over spirits through the development of techniques—but not to replace them with a new system of explanation. *Protoscientists*[2] and inventors of the time created new models and new things, but the construction of new explanatory models was retarded by beliefs in established models.

Philosophers sought to explain the occult world in terms of emotions, needs, and hierarchy. In an age in which complex machines were rare, they looked to natural systems for metaphors to explain their world. Language (in the form of spells, rites, and incantations) and mathematics were thought by them to be ways to predict and control fate.

The ancients searched the skies for omens of human destiny. Their search for patterns in the sky led to the development of the first predictive models of astronomy, but these were initially explained through astrology.

Mathematical models developed in the search for mystical relationships governing physical forms. To be sure, mathematicians developed descriptive models in physics by applying mathematics to concrete phenomena, but their explanations for cause-and-effect relationships were often mystical. Pythagoras (~500 BCE), for example, established a cult that pursued mathematical knowledge as a way to understand the world of ideas, not the physical world. The belief that numbers and mathematical forms have mystical powers is known today as numerology.[3]

Medicine developed along the model of technology, not science. Protoscientific models that we know of today are largely descriptive, mathematical, and logical—not experimental. Democritus (450–370 BCE), for example, proposed the existence of atoms in a void but he arrived at his model through reasoning alone, without experimental facts to back him up.

Likewise, Aristotle, who is known today for his direct investigations of nature, did few, if any, experiments. His ultimate goal was to understand the hidden world of ideas (i.e., the ideal world of forms thought to be the universal template for all existence) by studying the natural world. His models were largely descriptive and analogical—a curious mix of precision and assumption. For example, he attributed living will to inanimate substances as well as to animate beings, reflecting the Platonic belief that all things have a soul. In his model, rocks fell to earth because they had an affinity for others of their kind. Gold and mercury attracted each other because of spiritual love between them.

These explanations were consistent with the metaphors of the time. The philosophers' lack of progress toward the mechanistic explanations of modern science was not due to lack of logic, intelligence, or a desire to know; rather, it was due to the nature of the existing explanatory models. The same pattern typified intellectual inquiry in China, India, and, later, the Muslim worlds.

---

2   Referring to studies that would eventually give rise to science but did not have the method or purpose of science, such as astrology and alchemy.
3   Representing the Devil with the number 666 and viewing the number 13 as unlucky are examples of numerological beliefs.

True, individual scholars sometimes inquired systematically and even did experiments. In the 11th century, Ibn-al-Haytham (CE 965–1039), also known as Alhazen, espoused experimental ideas, but he did not foster a scientific tradition in the modern mold. Alhazen's work appeared in Europe with the rediscovery and translation of ancient and Islamic texts by European scholars in the 13th and 14th centuries.

Roger Bacon (~1250) and others also proposed elements of experimental science, but the movement toward systematic science did not pick up steam until the 17th century in Europe. Sir Francis Bacon, in his book *Novum Organum* (1620), laid the groundwork for modern science by championing *induction*—the idea of building models up from facts. European thought had reached a tipping point by then. The old explanations no longer satisfied them. A combination of advancing technology, affordable books and paper, discontent with scholasticism,[4] wealth from the new world, growth of mass production, a spirit of exploration, deism,[5] and a growing demand for individual rights was fueling a new age of invention and exploration. A philosophical movement toward humanism, especially in northern Europe, gave birth to a new kind of intellectual democracy: an experimental new science in which anyone with the time, talent, and resources could participate.

> **Stop a Minute**
>
> **TEXT BOX 3.1**
>
> Timelines for science and technology are readily available online by searching "technology timelines" and "science timelines." Download one of each and study them carefully. Notice (a) the kinds of inventions or scientific models created (descriptive, explanatory); (b) the discipline (math, astronomy, physics, etc.); and (c) the cultural context (Greek, Roman, Arabic). These shift over time. What sorts of patterns can you find?

European natural philosophers did not reject occultism immediately. A number of well-known 16th- and 17th-century philosophers such as Newton, Kepler, and Boyle (as well as the Bacons) dabbled in astrology and alchemy. Their explanatory models referenced spirits and God, but as mechanical explanations developed and proved effective, the reliance on mystical explanations decreased. Metaphorical explanations alone lost their power.

The greatest shifts away from mystical model building took place in the 18th, 19th, and 20th centuries, with advances in instrumentation and methods, statistics, and the growing acceptance of a mechanistic worldview.

Although science as we know it could not have advanced without technological invention, early science had little impact on technology. Masters of the technical trades and the inventors of the 16th and 17th centuries were often educated as apprentices. They had little use for books on science. Technical books and manuals were available and took priority for them.

---

4   Argument based on preconceived religious and classical models.
5   A belief that God exists but is not involved in running the world.

As books became more affordable and science gained ground in the popular imagination, engineers and inventors began to read science and found explanations that informed their development of new techniques. A symbiotic relationship of sorts developed. Science and technology became inseparable companions and eventually, they became wedded in the popular imagination.[6]

Rethinking the Relationship of Science and Technology

This historical development represents a change in human thinking: one that is by no means complete in the 21st century. Keep in mind that the terms and perhaps the concepts of science and technology came into use in relatively recent times. The term *science* came into use earlier, around 1300 CE, at about the same time that Europeans were establishing the first universities, but it denoted any recognized area of scholarly learning including astrology and alchemy.

The "new science" of Bacon began to coalesce in the 17th century, while the use of the term in the modern sense, based on particular methods of investigation and model building appeared only around 1725. William Whewell is credited with first use of the term *scientist* to refer to one who practices the new science in 1834.

The term *technology* came into use between 1605 and 1615, despite the fact that human technology dates back millions of years. In a real sense, people prior to these times may not have conceived of their activities in the same way as we do today, because they conceived of the world in a much different way than most of us do today.

The relationship between science and technology is entangled. Table 3.1 outlines some of the usual distinctions drawn between them, but these distinctions are not so clear in practice. For example, applied scientists may add to theory while developing new techniques and creating products. Science is itself a set of techniques for building mental and expressed models. So why is science considered by some not to be a technological endeavor?

The answer may seem at first glance to be excessively particular, but the distinction is important. Scientists may well create techniques to achieve an end, but that end is the creation of descriptive and explanatory (theoretical) mental models.

Technology, in contrast, is the study and development of techniques in the practical arts for the sake of solving problems in those arts. This is a form of mental models building too, of course, but as we pointed out in the last section, finding techniques is not the same as understanding and explaining them. That this distinction may be challenged by some philosophers speaks to the difficulty of pinning down the meanings of the mental models represented by the words in our language. But the distinction can be important for this reason: Most modern democracies recognize the right of free speech and free expression of explanatory mental models. In such societies, ideas are commonly viewed in law as morally

---

[6]   Science, in the form of astronomy and physics, was introduced into the Harvard curriculum only in 1701, and was not given much play.

TABLE
3.1

	Science	Technology
Comparison of Science to Technology		
Mission is ...	Finding causal relationships in the field of study	Finding ways to achieve, or better achieve, a given end
Focus is primarily on ...	Construction of predictive, explanatory models: on knowing	Creation or improvement of useful products: on design and innovation
Processes center around ...	Systematic description, modeling, prediction and testing of relationships for explanation	Systematic testing of systems, materials, prototype models, and real systems for products
Products are primarily ...	Research papers and presentations	Practical products and technical papers
Success is achieved when ...	Models are accepted by scientific peers	Products work as desired
Ethics promote ...	Models free of extraneous political, religious, economic, and social influences	Products shaped to meet political, economic, social, and other practical ends
Interrelationship	Uses technology to extend the senses into new realms and to test ideas	Uses science as a source for understanding and exploiting processes of interest

neutral. You can hold any ideas you want to—even share them—as long as you do not act on them.[7]

Therefore, the descriptive and explanatory models of science have no innate moral or ethical value: they are simply models of what we believe exists. We make them good or bad when we must use them to create practical techniques for action. The explanation of how to clone cells (science) is different from creating techniques for cloning humans (technology), though they may be intertwined.

Maintaining a separation between the concepts of science and technology is one way of protect science from unwarranted external control. The ultimate deciding factor as to whether a project is scientific or technological rests with its purpose: If the predominant goal is to construct a descriptive or explanatory model that adds to a larger theoretical model, then it is science. If the predominant purpose is to develop a technique with practical consequences, then we may assign the activity to the domain of technology. Table 3.2 may help clarify this distinction.

---

[7]   In practice this is not always the case, of course; but the principle is important to scientists.

TABLE 3.2

Illustrations of Jobs in Science as Opposed to Technology

Job	Description	Application	Area
Cosmologist	Study of origin and evolution of the cosmos	Model cosmological evolution	Theoretical science
Population geneticist	Studies mechanisms of mutation in populations of humans	Understand past events by studying gene mutations in humans	Theoretical science
Wildlife biologist	Studies elk population to manage and maintain state herd using scientific techniques	Manage hunting in specific herd (no theoretical implications)	Technical/applied science
Medical researcher	Studies mitochondrial evolution in search of specific tags for a specific disease	Predict disease in a genetic line and also add to theory of mitochondrial evolution	Theoretical science/ medical technical hybrid
Medical instrumentation engineer	Create more effective stent for use in treatment of liver disease	Manufacture of better stent that may help save lives	Technical/invention

The Processes of Science

Scientists create models that describe and explain natural phenomena. *Natural phenomena* are events of the everyday world that can be witnessed either directly or by measurable and replicable effects. The process begins, as all learning does, with the creation of a mental model, which is then expressed externally for communication.

We commonly call this process *discovery*—meaning literally to uncover and reveal something. Discovery is necessary, but not sufficient to explain science. Once we discover a phenomenon, we first have to describe it accurately; then we have to propose and test explanations for it. This is an act of creation, not just a process of finding something. The thing we create is an *expressed (mental) model*.

The descriptive and explanatory models accepted in science must ultimately be grounded in fact. *Fact* in science refers to a detectable event that we can record as *data*. We use facts to substantiate conclusions, or *inferences*, about events. The distinction between facts and inferences is important because people often state inferences as facts, when (in fact) they are not. Inferences may be *factual*, (i.e., based on fact) but they are not facts. If I touch a piece of fruit and find that it is soft (a fact), I may infer that it is ripe (or overripe). It is not a fact that it is ripe. Ripeness is not an observable quality.

Ideally, the scientist who examines an unfamiliar phenomenon first records facts. From these facts, she constructs a descriptive model using whatever tools she feels is most appropriate. Her next step is to compare her descriptive model to existing models, looking for patterns and analogies that would help her to explain the new phenomenon.

TEXT BOX 3.2

Darwin and Evolutionary Theory

The story of Charles Darwin, his voyage on the HMS *Beagle,* and his subsequent influence on evolutionary theory is well known. We can use Darwin as the centerpiece as we look at how a theoretical model develops. Darwin did not originate the theory of evolution, as some people suppose. Elements of the theory go clear back to ancient Greece, where some philosophers (Anaximander and Empedocles) suggested the fossils and ancient bones they found were evidence of an evolving Earth of great antiquity. The basic ideas underlying modern evolutionary theory existed before Darwin took his voyage: that is, descriptive and speculative models existed. The explanatory component—the mechanism driving evolution—was missing.

In the course of the voyage, Darwin collected specimens, adding to the database he would later use to build his models. It was, as is famously known, Thomas Malthus who suggested that competition was a key component in limiting human populations. Darwin (and Wallace, who developed the same idea) saw in Malthus's idea the key to a theory of natural selection. The natural selection model he developed was important because he supported it with a convincing descriptive model. Darwin and Wallace had added the element that turned the speculative notion of evolution into a true theoretical model.

The inductively created and largely speculative model of evolution now had an explanatory heart. Even so, it might not have gone anywhere without further testing against the predictions it now afforded. Darwin and other scientists were able to make and test predictions of what they might find if the theory of evolution and natural selection were true. The addition of a fully consistent theoretical model of genetics further supported and affirmed the model of evolution we now accept. Evolutionary theory today is a mature model fully accepted today by reputable scientists in the field.

The logical process of creating an explanatory model from specific facts is called *induction*. The speculative model that results from induction is a *hypothetical model* (or hypothesis). However convincing such a model seems on its face, it is considered a tentative explanation until it is tested. If more than one model suggests itself, you must test the competing models.

There's nothing magical or mysterious about hypothetical models. We create and test them whenever we have a problem to solve. Suppose, for example, I walk out one cold winter's morning (always in winter) and find my car battery completely dead. Knowing something about cars, I consider several explanations: (a) the battery died of old age; (b) I left something turned on and drained the battery; (c) the alternator isn't charging the battery, or (d) I have bad wiring that's draining the battery. Each model is testable, as shown on Table 3.3. Testing allows me to find the true problem so that I don't waste time and money on repairs I don't need.

It might be tempting to accept a hypothetical model just because it is intuitively simple and attractive. I might run out and buy a new battery because this is a quick solution and—after all—the battery is four years old. But if I am smart and patient, I will test my alternative hypothetical models systematically. Now I'm working from a series of models toward a specific conclusion. I can make predictions, in turn, about what I will find if each given model is correct. Testing a model by testing predictions made from it to arrive at a specific conclusion is a process called *deduction*.

TABLE 3.3

Model for Hypothetical Problem Solving

Hypothesis	Test	Prediction if Correct
Battery has given out with age.	Charge with a battery charger.	Battery will not hold a charge or charge will be very weak.
I left something on that drained the battery.	Charge battery and search all switches and doors.	Battery will charge. I will find the radio or lights left on.
The alternator is not charging the battery or is charging at too low a rate to replace charge.	Run engine and test current at the battery terminals.	No charge or a weak charge will register at the battery terminals.
Bad wiring is draining and discharging the battery.	Charge battery and test for high battery discharge while the engine if off.	Battery will charge. Discharge will be above the normal level for an idle car.

The one-two combination of induction and deduction is known formally as the *hypothetico-deductive method* (Figure 3.1). William Whewell, in 1837, called it the *scientific method*. Scientists use other methods to build models as well, but the hypothetico-deductive approach has played an important role in the development of science.

The Design of Scientific Tests

Scientific models must stand up to the scrutiny of professional peers,[8] just as the model of a criminal case created by the police and prosecutors must stand up to inspection by a jury. Because these experts are aware of acceptable methods and theory, they are more critical and less susceptible to emotional persuasion than most criminal juries are. A scientist's reputation depends upon his or her ability to put together expressed models in the form of research papers and presentations that are convincing to peers.

What are the parts of a scientific test? The important elements of all models are parts and relationships. The systems portrayed in scientific models consist of variables and constants. *Variables* are parts that can change in value. *Constants* do not change. Your heart rate is a variable. The speed of light in a vacuum is a constant.[9]

Scientific tests are usually designed to find out whether changes in variable A cause changes in variable B in a regular and predictable way. In this case, we would change the value of A and measure the resulting changes in B.

Because we are manipulating variable A, it is called the manipulated, or *independent*, variable. Variable B responds to or depends on changes in A, so it is called the responding, or *dependent*, variable. And, since variables A and B are usually embedded in a larger system, we also have to take into account *extraneous variables*—let's call them C and D. C and D may influence B and so cause or inhibit the observed effect on B, so variables C and D have to be either accounted for or controlled.

The best way to *control* C and D is by holding them constant during the test. If they don't change and B does, you can assume the effect is due to variable A.

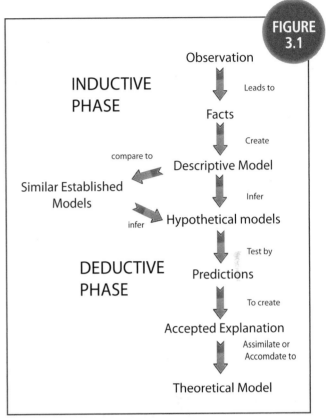

FIGURE 3.1

This model for creating theoretical models shows the inductve and deductive stages. The model is linear and idealized. Real science is not quite so neat.

---

8   Experts in the same specialized discipline
9   The speed of light does change in different mediums.

If you can't hold them constant, you can account for C and D by setting up a *control test*.[10] A control test is run in exactly the same way as your treatment test, except that you don't manipulate A. In other words, you allow C and D to operate in both tests. If B changes in the experimental test but not the control test, you can assume the effect is due to A.

Let's say we want to create a model of the influence of potassium on plant growth. We grow three dozen plants and treat them with varying levels of potassium fertilizer (the manipulated variable) including no fertilizer (the control). You expect potassium to influence their heights after three weeks (the dependent, or measured, variable). Height may also have something to do with the amount of water and sunlight intensity the plants receive, so we give all of the plants exactly the same amount of water and sunlight. We also use seeds from the same packet and soil from the same source for all of our tests. Assuming that we have identified and controlled the important relevant variables, we should be able to create a reasonably good model from this setup.

On occasion, we may suspect that variable A causes changes in variable B, but we can't manipulate A. A medical team, for example, might suspect that a particular chemical (A) causes birth defects (B) in children, but it would be unethical to expose healthy children to chemical A.

Instead, you test your model by determining the incidence of birth defects (B) among two groups of children, one group consisting of children exposed to chemical A and the other consisting of children not so exposed, but who are otherwise similar in gender, age, weight, ethnic background and other variables needing control. If the children are matched well enough, you can attribute any major differences in the incidence of birth defects to chemical A. The problem with this approach lies in getting a good match between the groups.

In a third situation, you might not be able to manipulate or compare groups. Events such as the origins of the cosmos or the evolution of life forms are long-term, one-time events. To test models of such events, scientists look for independently constructed models that verify the facts and relationships in the target model. In the case of our model of biological evolution, for example, we find that the model is supported by independently constructed, factually verifiable models in genetics, embryology, paleontology, and microevolution.

As this section illustrates, there is actually no one scientific way to build a model. Scientists adapt their methods to the problem. But every scientific model is an argument: a case being made in the court of scientific experts.

Quantitative and Qualitative Models

We have already seen that science has long been intertwined with mathematics. Copernicus, Alhazen, Newton, and Galileo studied and taught mathematics. Gauss called mathematics the "queen of sciences." Some philosophers of science

---

[10] Control is both a noun and a verb. A control as a noun is a comparison—a control test. As a verb, it means what it says—exerting control over a variable or situation.

today do not consider math a science in the modern sense because it does not rely on experimental methods, but it is still essential to most scientific models.

Because numbers are more descriptively precise than words, many natural scientists have a strong preference for building models using mathematical (*quantitative*) data. Scientists in fields such as cosmology and quantum physics may work almost exclusively with mathematical models. So how can we accept a mathematical model as scientific when it contains no empirical facts?

As we noted earlier, all explanatory models in science begin as hypothetical models. Speculative (hypothetical) models must be affirmed though testing. These models are "scientific" to the extent that they are consistent with empirically verifiable models that are already accepted by the scientific community. However, even the most intuitively logical model may prove to be wrong in the end. An untested model based only on logic, mathematical or otherwise, will only gain so much support from the scientific community. A purely mathematical model will not survive unless it can predict some detectable phenomenon.

Of course, scientific models can also include nonnumerical data. Because these data describe the qualities rather than quantities of events, they are labeled *qualitative data*. Qualitative data includes any records of facts in nonnumerical form, such as written descriptions, photographs, drawings, and so forth.

Linnaeus relied upon qualitative features of living things when he constructed his system for classifying plants. Darwin and Wallace constructed their models of natural selection primarily from qualitative data. Qualitative data are necessary in order to make sense of quantitative relationships (some quality must be measured).

Most models constructed in the early stages of a science are qualitative. At that point, scientists are searching for testable relationships. Once they are clear on what relationships to test (as clear as they can be), they turn to numbers.

One problem with purely qualitative models is their imprecision: the difference between saying that "*most* X are Y," and "63% of X are Y." In addition, we can test large amounts of quantitative data statistically, whereas large amounts of qualitative data are difficult to assess and evaluate. On the other hand, the use of statistics does not necessarily give a model credibility. Most of us are familiar with the phrase "lies, damned lies, and statistics." There is a reason why the cynicism exists. We will look at this problem in another chapter.

Statistical Reliability and Validity

How do statistics work? Think about flipping a coin. For a normal coin, on average, there should be no statistical difference between the number of times it comes up heads and the number of times it comes up tails: the ratio should be one head for every tail, expressed one head to one tail, or 1H:1T.

But if you flip the coin only six times, chances are you would get a much different ratio, say 4H:2T. The more times you flip the coin, the closer you will come to a ratio that can be reduced to the expected 1:1 ratio.

Make sense so far? Now let's say, we flipped a coin 30 times and came out with a ratio of 1H:1.1T. We flipped again 30 times and the ratio is 1.2H:1T. A third time we get 1H:1T. We continue to perform this test many, many times. Eventually I can construct a model like the one in Figure 3.2 that shows the range of ratios that I can expect from a fair coin flipped thirty times, and about how often I can expect each ratio to happen.

FIGURE 3.2

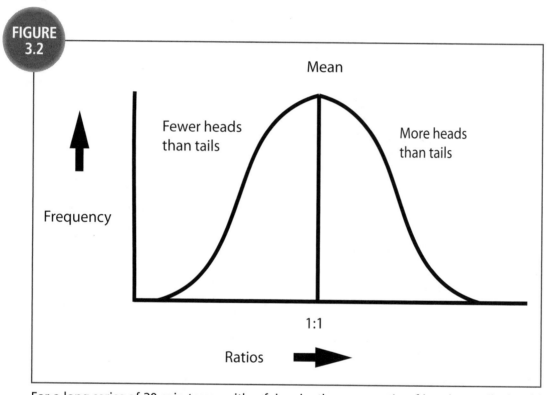

For a long series of 30 coin tosses with a fair coin, the mean ratio of heads to tails should average to a 1:1 ratio. The further from the mean a ratio is, the less probable it is.

Now suppose that I suspect that a coin is unfairly weighted so it will come up heads more often than tails. If it is *not* weighted, the ratio of heads to tails after a large series of 30 flips should fall around 1:1. If it is weighted, the ratio will favor heads over tails. The farther off the expected 1:1 ratio the results are, the more likely it is that the difference is not just due to chance, but that there is a real difference between the head and the tail side of the coin.

That is a nutshell overview of how the most common statistics work. Remember, in the tests we discussed earlier, we are looking for differences in a variable we measured in two groups. By comparing the actual means for the groups to the mathematical means that you might expect if there was no difference, you can calculate a probability that there is a real difference between them.

This assumes that your data have a certain kind of natural distribution. Mathematicians assume that the measured values of any single physical characteristic in a large population will show up as a normal curve when they are graphed. The center line of the curve is the mean value for the characteristic.[11]

If you measured and graphed the hand widths of all women in a large population, you would end up with a curve similar to the one shown in Figure 3.3. This figure represents a *normal* or *bell curve*: most hand widths cluster around a mean for the population. There are fewer really big or really small hands.

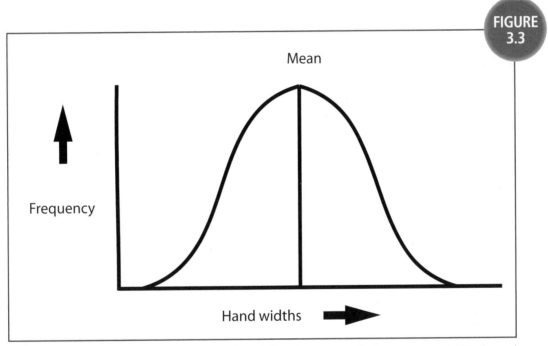

FIGURE
3.3

If you measured the width of women's hands in a large population, you would likely end up with a distribution like this one. It is called a normal or bell curve. It may be narrower or wider, but is typical of the range of values you get from many independent measurements of a phenomenon.

The bell curve is a standard model, but it may be high and narrow or low and wide, depending on the characteristic. If you collect a large number of independent measurements of a thing and the measurements cluster tightly around the mean, your data are reliable. *Reliability* is a measure of how likely it is that your calculated average is correct. We can illustrate reliability metaphorically by

---

[11] Mean, median, and mode (the three statistical "averages") will all be the same on a perfect normal curve.

using a target-shooting analogy. If you were to fix a gun in place and fire it at the center of a target, the clustering of the bullet holes around the same spot would indicate the reliability of the gun. A reliable gun will send bullets very near to the same spot on the target every time it is fired. If the holes are widely scattered, the gun is unreliable.

Validity, on the other hand, refers to whether the data you collect is actually a good measure of what you think they are measuring. Sometimes we can collect highly reliable data, but our data don't represent what we think they do. For example, we may collect data on reliable body weight in a population to indicate its general state of health, but unless I have shown there is a predictive relationship between body weight and general health, the validity of my conclusions might be questioned.

Good validity and reliability are essential in scientific model building.

Scientific Models and the Stability of Targets

The success of science generally depends upon the stability of the phenomena—the target—being studied. The natural sciences tend to model stable targets, while the social sciences deal with much more variable phenomena.

That physics and astronomy evolved first among the natural sciences is no accident. The reliability and predictivity of the models created in these areas are usually pretty stable. If you exert a particular force on a lever of a certain length at sea level, you can lift a predictable weight with a predictable advantage. Stars appear in regular patterns and move in predictable paths—for the most part. This consistency not only had practical value, but it also likely appealed to the aesthetic sense of the early natural philosophers and mathematicians. Mathematics was seen as a universal constant, and consistency was a valued commodity in a tumultuous world.

Biologists, on the other hand, have always had to deal with considerably more uncertainty. The widely cited *Harvard Law of Animal Behavior*, for example, states tongue-in-cheek that, "under carefully controlled experimental conditions, lab animals will behave as they damned well please." The living world in general exhibited enormous variability and was more in the domains of the practical arts such as agriculture and medicine and less studied by mathematicians and philosophers seeking to unravel life's mysteries.

Meteorologists and climatologists still struggle in in domains exhibiting a great deal of variability. Even with powerful modern computers, satellites, and progressively larger data collection networks, weather prediction is still largely a matter of probabilities. Though it is likely that the ancients could recognize signs of good or bad weather, for the most part weather and climate were wild and unpredictable: a product of the will of the gods.

No matter how precise a science may seem in its predictions, though, all scientific models are based on *probability*: the likelihood that an event or behavior will occur. The universe is not a stable place. On the whole, it is becoming more disorganized (measured as *entropy*) even as some systems (such as living organisms) become more organized (a process sometimes called syntropy or negentropy). The negentropic forces creating stars, planets and life are considered temporary conditions in the evolution of the universe. They are constantly opposed by entropic forces; thus, natural systems and their components are always in flux.

Simplification and *error* are inevitable occurrences in science, just as they are in all learning. In real science, just as in school science, models don't always turn out the way we want them to. Extraneous variables interfere with our efforts to construct accurate and predictive mental models. Out of ignorance or carelessness, we may us inappropriate methods, choose the wrong variables to measure or control, process the data in the wrong way, or misinterpret the outcomes we get. Our existing mental models may bias our interpretation of our new model. Such problems crop up in all mental model building, scientific or otherwise.

As we have already seen, no mental models are objective; they are, rather, subjectively constructed descriptions or explanations that we develop out of the materials available to us: our subjective perceptions of reality and the reference models we already have. We can never know if a model is true under all possible circumstances, but if they work to explain the circumstances that we normally encounter, we accept them as true.

Parsimony and Model Building

In some cases, two or more competing explanatory models may seem equally well supported by accepted facts and assumptions. In deciding which model to accept, scientists may consider each model's simplicity.

Just as juries are more prone to accept an explanation that is simpler over one that is complex, all other things being equal, so scientists are inclined to favor models that are simpler and more plausible in terms of its testable assumptions, its logical inferences, and its consistency with existing theoretical models.

A simple way to state this principle of *parsimony* is that, given two or more competing and equally plausible explanations, the simplest one is most likely correct. This principle, also called Occam's razor, originated in the 14th century with William of Occam, and has been affirmed as a principle of science ever since.

An MBST View of Modern Science

Scientists create mental models and share them as expressed models, primarily in the form of research reports and conference presentations. In the early days of modern science, these reports often took the form of letters to a recognized group such as the Royal Society or the French Academy, the readings of which often included demonstrations.

Scientists today usually present their models at scientific conferences or publish them in scientific journals. Most of their presentations and reports are *peer-reviewed*, meaning that editors and conference chairs send them out to experts in the field for review before accepting them. Reviewers may then recommend acceptance, rejection, or revision, if they don't believe the reports meet the standards for quality accepted in that field.

In theory, peer reviewers don't reject the work of colleagues merely because they don't agree with their models; rather, they look primarily at the technical merit of the work, such as

- internal consistency, including logical and transparent methods, predictions (when appropriate), data, analyses, and conclusions;

- reliable empirical data and appropriate methods for analyses of the data; and

- conclusions that follow from the data and add to appropriate theoretical models.

Although it is sometimes misused and abused, the concept of peer review is extremely important for maintaining integrity of science. Critics who argue for the publication of certain speculative or nonscientific models sometimes mischaracterize peer review as a kind of censorship. In their view, science is too unreceptive to models that contradict established theory and thus inhibits the very progress it claims to foster.

The history of science, though, establishes that such professional censorship in science is rare and self-defeating. It is true that scientists with an established theoretical viewpoint may resist changing them. That is human nature. We protect that in which we invest. But new scientists in particular are always looking to create new and fruitful models upon which they can build their reputations. They will not long reject promising new models.

This *competitive nature of science* is an important safeguard against prolonged scientific bias favoring a model that has passed its prime. Scientists build their reputations and acquire the rewards of success through their demonstrated abilities to construct models consistent with the physical evidence. The most valued models are those that fit their target, but then lead to new questions and opportunities for research and model building. If a model is objectively wrong—not true to the facts—it will be unproductive. The peer review system is one way to ensure that scientists are not accepting models that would confuse and impede their pursuit of fruitful explanations.

Template for the Scientific Model

Most scientific models follow a similar format, although sections of the models may vary in labeling according to the preferences of journal editors. In general, they include

- a statement of the problem;

- a justification clarifying the need for the model;

- a precise description of the methods used to build the model;

- the hypothetical models being tested, if appropriate;

- a data analyses and statement of results in relation to the stated problem; and

- a conclusion relating the model to a broader theoretical model.

A descriptive model will differ in form from a model including hypothetical testing.

The form of the scientific model has evolved through tradition. There is no single governing body who defines what science is or is not, or what its models should look like. The organization and practices of science have evolved because they work, and the form of these expressed models reflects a way of communicating that has proven fruitful in the past.

The scientist is in essence making an argument for the correctness of his or her model, and by correctness, we mean that the model explains what we have observed referring only to natural forces. To be accepted by peers, a model must be testable, replicable, and falsifiable. It must be supported by data that can be verified, directly or indirectly. It generally must be consistent with existing theoretical models or must make a strong case that existing models are wrong. Finally, it cannot reference supernatural or undetectable and immeasurable forces.

Once a model is accepted, it becomes part of a comprehensive structure of scientific models that we refer to as *scientific knowledge*. Where do we keep these models? Who keeps track of what the theory of evolution or the theory of gravity looks like? Well, no one… and everyone. Scientific models, like all beliefs, are mental models first. We may find representations of them in textbooks, sourcebooks, general reading, journals, letters, page on the internet, and other sources, but each of these representations—each model—is somewhat different from the others, especially in the details.

Like our personal mental models, established scientific models contain core principles, laws,[12] and examples for a framework. But no one—even experts—can completely understand or describe all of the elements of a given theoretical model because every mental model is simplified and necessarily incomplete.

You may object that this means that nothing can be known absolutely, and you would be right. Human mental models are always in a state of change. My model of the world today is not the same as I had yesterday or will have tomorrow. But what I forget, or don't know about a thing, someone else may know. We store the information we need in computers and books (which are just inanimate extensions of our memories) and then assemble our mental models when they are needed. The

---

[12] *Laws*, in the parlance of science, are expressions of regular relationship without explanation. For example, the law of gravity is the relationship between two objects that we express as, "any two objects attract each other to their centers with a force proportional to the product of their masses and inversely proportional to the square of the distance between them," or "$F = G \, (m_1 m_2 / r^2)$".

important point is that no expert completely knows and understands the model in his or her field in the same way as other experts. That's why they sometimes disagree and argue over elements of the model, or they contradict one other.

We are constantly modifying our accepted scientific models, usually making them more complex by adding examples or making exceptions or building linkages to new models. Even foundational models such as the theory of evolution in biology or the theory of plate tectonics in geology are constantly being modified.

Specialization is the rule in science today. Few modern scientists have the ability to be polymaths—experts in a number of fields—like some of the natural philosophers of the seventeenth century. Just keeping up in a single discipline is often costly in time and resources; and as targets become increasingly remote and difficult to model, scientists must spend billions of dollars to build exotic machines such as CERN's particle accelerator, or the Hubble space telescope, or the labs on the International Space Station.

Summary

In the final chapter in this book, we will revisit certain elements of the nature of science that we have outlined here. My purpose in this chapter has been to construct a model of science as it is often described in the literature and to contextualize it within the framework of model building. This approach differs from traditional models of learning. Throughout most of human history, philosophers have sought to know truth with a capital T, which is to say, absolute Truth. Some of them took to the study of nature as a way to reveal such Truth. In the early days of the "new science," there were those who believed that mechanical philosophy, as it was sometimes called, could reveal firm and unchanging truth. There are, in fact, still some scientists and science writers who appear to believe their models represent unchanging universal truths. And this may be so; the problem is, they cannot prove their contention.

Scientific models, like all models, are simplified expressions of our mental models. Our individual models are likely never identical to one another, nor can they account for all of the possible relevant facts and relationships that may actually exist in the universe. This does not affect the value of these models, nor does it mean they are wrong, as far as they go. But we should think of our mental models as approximate representations of their targets.

Scientific models are true to the observed and agreed-upon facts. Because they have predictive and explanatory powers, they are useful. Some of them lead to useful technological applications. But they are models, and by definition, they cannot represent their targets fully. The only attributes of the targets that are available to us are those within the range of our physical capabilities: our abilities to observe and reason.

For Discussion and Practice

1. On the internet, search with the keywords "science timeline" and look at one of the timelines. In a paragraph, summarize evidence that technology developed well before science.

2. Is there evidence in the timeline that the development of mathematics was related to (or paralleled) the development of the protosciences?

3. Einstein said that scientific methods were merely extensions of everyday problem solving. If you had a troublesome decision to make, how could you use the hypothetico-deductive method to help you?

4. In your own mind at this point, is there any advantage to describing science as a process of model building rather than a way of discovering new knowledge? What is the difference in these two models of science?

References

Graf, F. 1999. *Magic in the ancient world*. Cambridge, MA: Harvard University Press.

Huxley, T. 1901. President's address to the British Association for the Advancement of Science, Liverpool Meeting, 14 Sept 1870. *The Scientific Memoirs of Thomas Henry Huxley*, 3: 580.

Kennett, D. J., and J. P. Kennett. 2006. Early state formation in southern Mesopotamia: Sea levels, shorelines, and climate change. *The Journal of Island and Coastal Archaeology* 1 (1): 67–99.

Melville, D. J. 2003. Archaic Mathematics. In *Third millennium mathematics*. *http://it.stlawu.edu/~dmelvill/mesomath/3Mill/archaic.html*

Plummer T. W., P. W. Ditchfield, L. C. Bishop, et al. 2009. Oldest evidence of toolmaking hominins in a grassland-dominated ecosystem. *PLoS ONE* 4 (9): e7199 DOI: 10.1371/journal.pone.0007199

Chapter 4

MODELS and SCIENCE TEACHING

All models are approximations. Essentially, all models are wrong, but some are useful. However, the approximate nature of the model must always be borne in mind…

—George Edward Pelham Box

We've spent some time now creating a model of science based in the concept of mental modeling. Hopefully, I've made the case that mental models are more than just imaginative metaphors. In order to know something, you have to build a mental model of it. We see evidence of this all the time, but we have not recognized its importance as a key to understanding learning.

The example of Models-Based Science Teaching (MBST) I present in this chapter is built upon the guided inquiry model. Inquiry has shown itself to be at least as effective as traditional approaches for content acquisition and better in creating positive student attitudes toward science. On its face, inquiry also better mirrors the practice of science than traditional approaches. Our purpose is to impose a framework that develops other elements of science literacy as well, specifically understanding of the nature of science and its personal and social contexts. Building upon inquiry makes good sense, but you could infuse elements of MBST into a more traditional curriculum.

MBST is largely a matter of discourse—of how you present and explain the components of science to your students. It does entail learning a new set of skills, but these skills are readily learned by anyone who accepts the premises of MBST. Ideally, the language of MBST would suffuse the science curriculum from elementary school through college, but even if it doesn't—if it is used only in your classroom—it can do some good.

Content to Inquiry to Model Building

The National Science Education Standards define inquiry in science teaching as:

> . . . a multifaceted activity that involves making observations; posing questions; examining books and other sources of information to see what is already known; planning investigations; reviewing what is already known in light of experimental evidence; using tools to gather, analyze, and interpret data; proposing answers, explanations, and predictions; and communicating the results. Inquiry requires identification of assumptions, use of critical and logical thinking, and consideration of alternative explanations. (NRC 1996, p. 23)

Skill in inquiry is one of the four important dimensions of science literacy referenced in the Standards, along with learning of science content ("subject matter"), understanding of the nature of science, and ability to view science in a greater social and personal context.

Inquiry differs from more traditional lab approaches in a number of ways. In the traditional lab activity, the focus is more heavily upon learning the content. Processes of investigation tend to be limited to manipulative and data collection skills (Table 4.1). Lab and field activities are usually prescriptive (sometimes called "cookbook labs") and are intended to lead students to correctly understand the phenomenon under study, which has usually already been discussed in class. For the most part, the students are not creatively involved in the development of the activity. Often they simply fill in the blanks on a printed sheet to complete their lab.

Inquiry, in contrast, focuses more on the development of the scientific process skills, as shown in Table 4.2. Content is developed with a conscious focus on developing a conceptual network, rather than being given as information. Guided inquiry, which involves teacher participation, is most common in schools.

In guided inquiry, students examine a new phenomenon (often in the form of a problem) before the teacher discusses the underlying concept with them, although the teacher usually involves the student in the design of the inquiry and interpretation of the results. The amount of guidance provided by the teacher varies with the subject and the age and experience of the students.

TABLE
4.1

Traditional vs. Inquiry Curricula

Traditional	Inquiry
Content comes before activity	Activity leads to content
Content is information	Content is concept-focused
Activities confirm content	Activities lead to discovery of content
Little student involvement in planning	Some or much student involvement in planning
Heavily teacher directed	Some teacher direction
Focuses on obtaining answers	Focuses in part on processes
Emphasis on developing physical techniques	Emphasis on exploration and inquiry
Saying little about nature of science	Partly mirrors technical nature of science
Provides little context for science	Provides some context for science

TABLE
4.2

Basic and Integrated Process Skills of Science

Basic science process skills:	Observing
	Measuring
	Inferring
	Classifying
	Predicting
	Communicating
Integrated science process skills:	Formulating Hypotheses
	Identifying Variables
	Defining Variables Operationally
	Describing Relationships Between Variables
	Designing Investigations
	Experimenting
	Acquiring Data
	Organizing Data in Tables and Graphs
	Analyzing Investigations and Their Data
	Understanding Cause-and-Effect Relationships
	Formulating Models

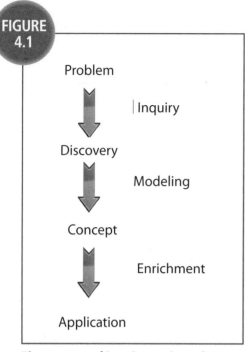

FIGURE 4.1

Problem

| Inquiry

Discovery

Modeling

Concept

Enrichment

Application

The process of inquiry as viewed through the MBST framework.

In contrast with the traditional approach, in which the activity follows the introduction of the target content, inquiry leads the students to the target from the initial activity. This means that students develop a mental model of the target concept *before* the teacher labels it. In the MBST format, the model for inquiry is similar to that shown in Figure 4.1.

Research to date indicates that students studying science through guided inquiry learn as much or slightly more science content than those following a more traditional curriculum and, importantly, report a greater liking for science. However, we have little evidence to show that students using inquiry gain a better understanding of the nature or context of science than those using a more traditional approach. This may in part be due to a lack of an *organizing framework*—a model that would provide inquiry with more meaningful form and focus.

All good therapists know how important a framework is. The same problem, seen through two different frameworks (some people prefer the metaphor of lenses), takes on different meanings. If we "do" activities without a framework, they are simply class assignments. They have no larger meaning. Part of the reason inquiry may not live up to its full potential is that the activities are often focused—like traditional labs—on content and may be too carefully guided by teachers. They do not involve students in anything much like science.

MBST provides you and your students with a thematic structure for inquiry. It does so by overtly defining science as the construction of descriptive and theoretical models. Building a model is different from "doing" a lab. It is different from inquiry alone.

A model is a *product* with a distinct purpose. We design models selectively, systematically, and deliberately. Building a model is an end in itself. You have a purpose for every model that you build; every sentence that you utter, for example, is a model.

Awareness of this process is the key to making science more meaningful and challenging for your students. And this awareness if often missing, right up through graduate school. Graduate students of science writing theses for their master's degrees or doctorates frequently have trouble with their research projects precisely because they do not have this end in mind. Stephen Covey's admonition to "begin with the end in mind" (1989) would be helpful. But what is the end? To build a model, of course. The ultimate goal of all learning and communications is to build an effective model. Learning how to build such models is in part what education is all about. Why do scientists build models? To describe and explain phenomenon.

In other words, science does not end at discover. It begins there. Inquiry is a means to an end, not an end in itself. Even if we keep our mental model to ourselves and never share it, we are still engaged in model building.

If we impart subject matter content and technical skills to our students without a firm grasp of the product, we will rob our students of a deeper understanding of science. I was reminded of the need for this kind of understanding when I described this models-based approach to a retired physician who had done science at one time as a grad student. I was stunned when he took vehement issue with the model I presented. When I pressed him to clarify what he was doing when he did science, he could go no further than to say he was discovering things. My effort to convince him that he was building a model failed. But then, I hadn't given him the background I've given you. Hopefully, you will be more receptive to the benefits of knowing about science as well as how to practice it!

Modeling Science at Different Grade Levels

What additions or changes do you have to make in your teaching in order to use MBST? The good news, if you use inquiry already, is not many. To make these changes, though, you must be thoroughly familiar with the concepts we have developed in the first three chapters.

The changes MBST requires are largely semantic—the product of a different way of conceptualizing science. Following from the axiom of beginning with the end in mind, what would a scientifically literate student know at the end of their student career if he or she was taught using MBST regularly and consistently in grades 4–12? The student would presumably

- know how to create descriptive or explanatory models of phenomena through inquiry;

- view science as the construction of descriptive and explanatory models;

- distinguish science from technology according to the purposes of each;

- understand learning as the creation of mental and expressed models;

- understand that knowledge is a variable structure of simplified mental models;

- understand the subjective nature of knowing and learning;

- know the difference between scientific and nonscientific model building; and

- understand on a practical level why science can never explain all things.

The principles underlying MBST are the same as those that frame our contemporary understanding of the nature of science and human thinking. As a teacher of science using models-based science teaching, you engage students in constructing a web of science concepts while also developing their understanding of

- how humans learn by constructing mental models (nature of learning);

- the nature and limitations of all mental models (nature of knowledge);

- the nature and limitations of expressed models (principles of communication);

- the processes for building models in science (scientific investigation and learning); and

- the differences between scientific and nonscientific models.

Clearly, these dimensions of MBST cannot be addressed in the same ways at all grades and ages, but let's examine how to adapt the method to each grade level.

MBST at the Elementary Level

Most elementary age children would be unable to comprehend models of the philosophical and scientific basis for human thought. They are just coming to grips with the nature of their own realities and have no idea how the brain works. Students at this age would not benefit from any but the most superficial references to the subjective nature of reality (e.g., "that's how you see it, but other people might see it differently").

However, elementary students can be given a broad definition of what a model is, and they can understand that they are creating models when they make something to represent something else. If they write a poem, for example, you can tell them they are creating a model of what they are thinking. If they draw and write a description of a leaf, you can remind them that they are creating a model in words and pictures of a leaf. You can introduce the word *target* to identify the thing they are describing or explaining. In this way, elementary teachers can lay the foundations for MBST and for understanding through model building.

As a simple example, suppose you as an elementary teacher want to engage your third graders in a study of insects. The goal of the exploration is to describe (create a model of) a typical insect. You give your students three pictures of insects and three pictures of noninsects (e.g., spiders and centipedes) and put your class in the role of scientists. You ask them to pick out the three and tell why they chose them.

Once the class agrees on the three candidates (with your help) you go on to help them build a model they could use to identify other members of the group called insects. The students could then compare the insects and noninsects and suggest features peculiar to insects to include in their model. While this seems like a simple activity, it actually mirrors the activities of early natural philosophers as they put together classification systems based on appearance.

It's not enough to stop there, though. The students will not understand anything about science by just constructing this model. You must explain to them that they, like scientists, are looking for patterns—common features the insects share that the noninsects don't have. You remind the class that they are creating a "model in their heads" of what an insect is and is not. When they write or draw the model on paper, they are doing what scientists do by creating a model to share with others from the ones in their heads.

Once the children have collected data, you work with them (or let them work together) to create the model. Once the model is created, they test it by looking at more insects, refining what they have created. Scientists, you tell them, have to test their models too.

You could do the same activity without MBST, but notice how the focus is shifted from content to building and testing a good model. That is the difference between having a framework and not having one.

MBST in elementary science focuses on

- broadening and developing the models concept to include mental and physical models,

- expanding the range of constructs that students view as models,

- engaging students in simple model building and testing based on existing inquiry activities, and

- framing activities with simple but valid ideas about what scientists do.

Keep in mind that all learning activities across subjects involve model building. When students do arithmetic, they are building models of the relationships of quantities. When they spell and write, they create symbolic models representing concepts and propositions. As an elementary teacher, you can apply the language of modeling across subjects.

MBST at the Secondary Level

By the time your students get to middle school, they could have a good grasp of models and modeling, assuming the exposure in elementary is similar to what we have just discussed. At this level, they should be ready to understand that

- scientists construct and share models of the natural world;

- scientific models are based on data that we can see and confirm;

- the data for a scientific model must be replicable and consistent;

- scientific models are what experts in an area agree upon;

- scientific models are accepted because they are useful and predictive;

- scientific models are simplified and do not describe or explain everything;

- scientists seek to know and explain;

- technologists invent and apply; and

- many of the models we commonly use are not scientific, but that does not mean they are wrong.

These ideas should grow naturally from your reframing of inquiry but, as we pointed out for the elementary grades, the ideas should be repeatedly included in your discussion of science with your students.

By high school, most students can think abstractly enough to handle deeper technical explanations of mental and scientific models. They should be mature enough to

- understand the theoretical basis for mental modeling and human thinking;

- understand science as a process of model building;

- understand the limits of theoretical models as both shared and individual constructs;

- design and construct scientific valid models of selected phenomena;

- use basic statistics or statistical presumptions to assess outcomes;

- critique scientific models as to their reliability and validity;

- critically relate their models to a range of targets;

- find new problems as the construct models;

- distinguish scientific from nonscientific and pseudoscientific models; and

- understand science in personal, social, cultural, and historical contexts.

While MBST may require expenditures of time to address some of these issues, such as the theoretical basis for mental modeling and human thinking, many of the principles of MBST can be infused into existing activities.

Since only one hour per week is devoted to lab work in the average high school classroom (NAS 2005), the approach should be infused into other modes of instruction. In biology, for example, you can talk about the Darwin/Wallace model (or theoretical model) rather than theory. You can refer to models of force, bonding, and plate tectonics and refer to factual models rather than calling them facts. And there are always opportunities in the course of instruction to talk about how scientists build and test models, and how the models are simplified, and to point out that not everything is known or certain. MBST requires a commitment to the development of scientific literacy. Without that commitment, it will go nowhere.

MODELS and SCIENCE TEACHING

Examples of MBST at Several Grade Levels

The steps you use to construct a scientific model are similar to doing inquiry activity, but with certain modifications. A full MBST-based inquiry usually includes the following steps:

- Introduce a problem of description or explanation
- Frame a solution to the problem in model-building terms by identifying the target
- Plan the model
- Construct the model by collecting and evaluating data and making inferences
- Evaluate the fit of the model to the target

In guided inquiry, the teacher usually poses the problem (open inquiry is more student focused). A *problem* is any situation or question for which we do not have an immediate solution, explanation, or answer. In the insect classification activity we discussed in the previous section, the problem was to identify the characteristics of the insects that distinguished them from noninsects.

Elementary students are usually more concerned with solving descriptive categorical and ordering problems than in constructing explanations for phenomena, although we should by no means exclude them from constructing simple explanatory models appropriate for their level of understanding and development.

By the middle grades (variously defined as grades 4–8), most students have the capacity for modeling concrete cause-and-effect relationships to explain a thing. For example, they can understand the causal explanation for winds and ocean currents by applying concrete convection models to these larger targets.

They are also able to understand abstract concepts such as density when the concept is presented in a concrete activity. In text box 4.1 is a classroom activity that uses differences among cereals to develop a mental model of density.

In this activity, the teacher provides the framework for the inquiry by suggesting a goal to the students. The problem is to explain why equal volumes of three different flake cereals weigh different amounts.

As we have pointed out, MBST depends upon regular and consistent use of models terminology. You never just "do" an activity without framing it in a models context:

Wrong: Today we are going to do an activity (or lab) investigating how the color of light influences the growth of bean plants.

Correct: Today we are going to practice science by developing a research-based model showing how the color of light influences the growth of bean plants.

TEXT BOX 4.1

The Density of Cereals

Purpose: To use differences in common cereals to develop the concept of density.

Students must be able to:

- Determine mass on a simple scale or balance

- Measure volume

Materials

- Three different flake cereals of distinctly different densities

Introduction
Put the students into the role of scientists who have never seen cereals. Their goal is to create a model to report back to their other scientists what they have found, providing as much information as they can. Ask what characteristics they might include in their models. Come to an agreement about what their models will include (making sure flake, compactness, weight of a cupful, texture, and so forth are included).

Procedure
- Give each group one cup of the cereal to work with.
- Allow them to collect the data agreed upon.
- Put numerical data on the board and calculate averages for the model.

Discussion

T: *Look at your data model carefully. What does it show you? What things that you say seem to be true?*

[They might see that the cereal with the smallest flakes weighs the most per cup. If they do, ask them to explain why this might be. Give them time to think. If they get stuck, guide them.]

T: *Do you see any relationship between the size of the flakes and the weight of the cup? They do. They state it correctly). Why do you think the cereal with the smaller flakes weighs more per cup?*

S: *'Cause you can get more flakes in. So there's more stuff in the cup.*

T: *More mass?*

T: Yeah, more mass.

T: *So there's more mass in the heavier cup than the lighter one? [They agree]. Weight comes from gravity acting on mass. You have the same amount of cereal by volume but different amounts by mass. So can you*

tell how much mass something has just by looking at it? [They say no]. *So let's say we know how much a cup of each of five cereals weighs. They look the same but they are different in weight—in mass per cupful. If you were given a cupful of one of the cereals but didn't know which one it was, how could you figure that out?*

S: *You could weigh it.*

T: *So you could use your model to figure out— to predict—what cereal you have. [They agree]. Scientists build predictive models just like the one you just made. They use the weight per cupful to help figure out what an unknown substance is. The part of your model that we've just discussed—mass per unit of volume—is a model of a concept called density. We would say these cereals have different densities because they have different masses in the same volume. You'll want to keep that model of density in your heads so we can add to it and make it grow.*

The script below imagines the teacher, who is overseeing the density lab described in the text box, guiding her students from that activity to another activity that will build upon the concept of density that the students have just learned.

T: *Okay, so we've created a neat scientific model of this thing called density. And if we knew the densities of all the cereals in the world, you could tell what kind of cereal you had just by measuring its weight per cup. Scientists like this kind of predictive model. It makes their job easier. Now we have this model in our heads we call density. It's basic, so we might want to make it better by enriching it. We've been talking about how much our cereal weighs per cup. What's wrong with using weight for our model?*

Ss: (blank looks)

T: *Is the <u>weight</u> of something always the same everywhere?*

S1: *No! You can be weightless in outer space.*

T: *But the amount of you doesn't change in outer space, does it? Do you remember what weight is?*

S2: *It's the pull of gravity on mass.*

T: *So the amount of material in the cereal is its mass. It stays the same no matter what it weighs?*

S1: *Unless you eat it!*

T: *Let's say we don't eat it. When you weighed the cereal, you found out how many grams it weighed because weight and mass are the same at sea level. So our weight in grams is okay as long as we are on the ground. But out model would be better if we used mass instead of weight. Agree?*

S2: *Okay.*

T: *How about our using a cup for the measurement? Is there any problem with that?*

Ss: (blank looks)

T: (takes out two cups of different sizes): *Here are two cups. Which one is the right size?*

S3: *I don't know. Either one could be right.*

T: *Exactly. To make a model that we can always use and understand with fewer mistakes, we will need a volume that always stays the same.*

S1: *Right.*

T: *Scientists want to create models that everyone can understand and duplicate if they want to. So even though we're okay with the grams per cup measure, we might want to use standard measures next time we build a* *model: measures that everyone knows. Now, here's an interesting question. You know there are different kinds of woods: oak, maple, pine, and so forth. Do you think they all have the same density or different densities?*

S1: *Different!*

S2: *The same!*

T: *What does the model we just created tell us about how we can find out?*

This dialogue, although obviously contrived, illustrates how you reference models and the nature of science in your conversations with students. You may focus on different aspects of science, of course.

Once the students have enriched their models of density by discovering that different woods have different densities, they could enrich their models by creating a link to the concept of buoyancy (denser woods float lower in water—are less buoyant—than less dense woods). If you are familiar with *learning cycles*, you will recognize the pattern. Let's pull another simple example from the field of chemistry, specifically endothermic reactions that absorb heat and feel cold as they occur, and exothermic reactions that emit heat and thus feel warm or hot.

In a traditional lesson plan, the concepts of endothermic and exothermic reactions are introduced to the students prior to them doing the lab. This is not so in guided inquiry. In the latter, students "discover" the differences before formally building a model by mixing the reactants and measuring temperature changes. The terms *endothermic* and *exothermic* are applied after they have constructed a mental image—a mental model—of the reactions.

Students can then enrich their model through a series of measurements of other reactants to create patterns that would predict endothermic and exothermic reactions.

This lab is a variation and extension of a typical "cookbook" lab, but poses a problem (to find a pattern in the model being created) and so is more interesting than most prescriptive lab activities. But MBST goes beyond inquiry. Having raised at least some interest and curiosity, the teacher assigns the students the role of scientists. The problem is stated after the "discovery," which is how to build a valid databased model of the phenomenon. By way of introducing the whole activity, the chemistry teacher might say something like this:

> *I've discovered an interesting phenomenon I'd like to share with you. It's so interesting that I think it deserves some investigation, but first I want you to see what it is for yourselves. This is what I did.*

The teacher then gives students minimal directions on what to do and they discover as-yet unnamed exothermic and endothermic reactions. After they have discussed the phenomena and named the reactions, the teacher continues, saying:

MODELS and SCIENCE TEACHING

Your job will be to work together as scientists to collect and pool your data to create a descriptive model of this phenomenon. I also want to see whether we can enrich our model by testing other similar materials to see whether they show the same or similar reactions. We will work out which materials to test in just a moment.

I also want you to think about <u>why</u> these differences appear. In other words, what could be going on inside the systems to cause this effect? These will be hypothetical models, of course, since they are speculative. But as we've said before, scientists often include their speculations in their descriptive models to suggest explanations for what they observe.

Since we will pool our data to help ensure its reliability, we first need to talk about, and agree upon, what you will include in your model.

In this brief introduction, you can see our emphasis upon building a model—not just completing an activity and finding right answers. Inquiry and discovery are part of the process but are a means to an end. That end is the final expressed model.

The teacher has laid out the parameters for the inquiry but has also left room for input from his students: that is, they plan the model together. In chemistry, of course, the teacher has to exert more control over most inquiries due to the inherent dangers in that subject. But students always know they are being guided in school science. Your goal is to use questions and suggestions to help them get a sense of how to shape a model—what variables to consider—even if you cannot give them freedom to inquire.

Almost all traditional labs can be transformed into opportunities for problem solving and model building. Although you may object that inquiry takes too much time, keep in mind that our goals have changed.

In the chemistry activity we just discussed, students will find that some chemical reactions absorb energy while others release it. They may hypothesize then that "all chemical reactions either absorb or release heat." When questioned about where the energy may come from or go, students may puzzle out the involvement of chemical bonds. This speculation can be included in their models *as speculation*. The postlab discussion might go something like this (in part) where the teacher uses the notion of the target:

T: *The targets for our model are the reactions that we actually investigated. But can we honestly say that all chemical reactions are either endothermic or exothermic?*

S1: *Sure, if the energy is in the bonds, it gets released. Or it takes energy to break a bond.*

T: *So what do we know for sure?*

S1: *If there's a reaction, bonds must be broken or formed.*

S2: (points to paper) *Or both. But all we know for sure is that the reactions we tested either released heat or took it in.*

T: *How reliable were our data?*

S3: *They were good.*

T: *How certain are we that our model includes all chemical reactions?*

S2: *Kinda certain. I mean, we can generalize from our model but we would have to look at all reactions if we wanted to be sure.*

T: *So we can't be sure from our model that all chemical reactions are either exothermic or endothermic?*

S1: *No, not really. Some we haven't seen might be different.*

T: *But it seems likely?*

S3: *Yeah.*

T: *So our model is useful but incomplete. Still, it's useful. And the explanation you proposed for why the reactions feel hot and cold is consistent with the existing scientific model for chemical bonding that's already part of our theoretical model.*

Ss: *Yeah.*

In guided inquiry, you provide the leadership needed for your students to make sense of their work. Students can enrich their models by searching out background information on endothermic/exothermic reactions once the concepts are introduced.

Now let's look at a third scenario, this time in biology. Our goal is to develop models describing and explaining adaptation in flowers. Our first step is to develop descriptive models of flowers. So we give our biology class several morphologically and ecologically distinct flowers and set up a problem using MBST terminology. The teacher says:

> *We've learned already that scientists create descriptive and explanatory models. Even today, there are scientists creating models of new species just as Ray, Linnaeus, Darwin, and others did several centuries ago. Genetics are used more now, but we usually rely first upon differences in form to create preliminary models.*
>
> *Naturalists have always being interested in creating models of why particular creatures have particular forms and functions: for example, why a leaf is shaped the way it is, or why bark on a bush has certain textures. Over time, they have created a theoretical model relating form and function.*
>
> *Here is the situation. Each team of naturalists has two newly discovered flowers. Now you are already familiar with flower anatomy from our work with the lily. Your task is to enrich your model of a flower by creating*

descriptive models of your two flowers, using the lily as a reference. In the end, you will want to be able to speculate on the causes of any differences you see among the flowers. Now let's consider what we might include in our models to make them useful. What should we do to make our models most meaningful and how should we record data?

Note in this monologue how the teacher tries to engage her students in role-playing. She gives them a situation and a task, which is to build a model. Of course, the students are well aware that they are not scientists, but they can still benefit by pretending to be. Role-playing can be a powerful way to develop their understanding of the nature of science, if you use it regularly and consistently, and are careful to model science well.

Part of your role as a teacher is to create a scene or role for your students. In many classrooms, that role is one of passive learner. In the best science classrooms, the learner is active and engaged, and has a sense of purpose. MBST is primarily about developing purpose.

Introducing Your Students to MBST

In the book, *Learning How to Learn*, Novak and Gowin (1984) stated their purpose as "educating people and… helping people learn to educate themselves. We want to help people get better control over the meanings that shape their lives." The authors felt that knowing how to learn is just as important as knowing what is learned.

MBST is likewise concerned with building a model of learning that helps students of all ages understand that when they learn something,

- they are constructing mental models that are by nature simplified and subject to change;

- these models are adopted because they work and not necessarily because they are the only true and most effective ways of understanding the world;

- no one has a complete grasp of any model, and most of the time we are working with approximations of a situation; and

- what we create when we communicate are expressions of our inner mental models alone.

These basic principles have implications that go well beyond our understanding of science. Learning how to learn (by building models) is in MBST supplemented by learning *what* we learn.

If you are an elementary- or middle-level science teacher, you will generally infuse MBST into your science activities without referring much to the underlying rationales we developed in Chapter 1 through Chapter 3. Young children are attuned to learning by doing. They would not understand such rationales. But if you are a high school teacher, you could more overtly include some of the

TABLE
4.3

Stereotypes of Science vs Alternative Tenets of Science

Stereotypes	Alternative Tenets
Science is a search for unchanging truth.	Science is a quest for predictive explanations.
Reality is objectively knowable.	Reality can only be known and understood subjectively.
Once something is proven, it doesn't change.	All scientific models are subject to change if change is warranted.
Scientists are inventors bent on controlling things.	Scientists build descriptive and theoretical models.
Scientists are objective and dispassionate.	Scientists think subjectively and are motivated by curiosity and desire for success.
Scientific knowledge is objectively proven reality.	Scientific knowledge is models accepted by the agreement of experts in a field.
Scientists are atheists opposed to religion.	Scientists need not be atheists but may reject some beliefs of established religions.
Knowledge is contained in books, journals, and other media.	Knowledge exists in the collective of human minds.

philosophical and conceptual ideas that underlie our modern understanding of science and learning. How you would introduce these ideas would depend upon the curriculum you use. If you design your own, you have more options than if you are forced into a lockstep with other teachers in your district. Even then, however, you can infuse many of these ideas into your explanations and discussions.

The model most adults have constructed of science contains many stereotypes held over from the recent and distant past. These stereotypical models need to be reexamined in light of developments in the late 20th and early 21st centuries, because they pop up regularly in the science reports of popular media (and even sometimes in professional literature). Some of these stereotypes and rebuttals are presented in Table 4.3.

Students need to think about what happens when we think. One way to introduce science to your (high school) students is to present them with examples of the products of science, such as copies of research papers from journals like *Nature* or *Science*.

In very short order, you can develop many of the ideas of science as model building; first, by considering what these papers represent and getting into the heads of the authors; second, by thinking back about how they were constructed.

This approach puts a human face on scientific model building. Although it is best to select articles that are reasonably transparent, transparency is not as important as the discussion about the nature of the papers themselves.

This approach is nothing more than an application of the principle of beginning with the end in mind. If students don't know the products of science, how can they be expected to understand what science is about? By examining real scientific models, they begin to understand how a scientist represents a target by using graphs, pictures, descriptive narrative, and so forth.

As students examine the papers, the idea that these models are simplifications of real systems can take hold. It also becomes easy for them to understand the limits of these models when they have the real thing before them; that models have to be interpreted to have meaning, and that what they have is a representation in symbols of some real target.

I am suggesting, then, that starting out the year by examining the nature of the products of science — research papers — in the greater context of how humans learn and know a thing is essential if we want our students to understand themselves and science better.

Lesson Planning for MBST

Good lesson planning is as essential to MBST as it is to any other approach to teaching and learning. If you don't know where you are going, you cannot know if you have arrived.

Each time you plan a class, you must ask yourself how the content can be presented within the framework we have developed. Keep in mind that you are not being asked to add new activities to your curriculum; rather, you are being prompted to create an explanatory framework for those activities that better describes the processes and products of human thought.

If we apply our MBST model to lesson planning, we find we can be effective by following certain simple rules and principles:

- Construct models, don't just give out information

- Be parsimonious: develop the most important elements of the model solidly

- Create familiar imagery to anchor propositional elements (develop pictures in the head to anchor spoken or written ideas)

- Link parts of the models and models themselves systematically and hierarchically

- Teach students how to create and evaluate scientific models through inquiry

- Create a model of what science is first by examining the products of science

- Engage your students in the role of the scientist as model-builder

- Always refer to the nature and context of science during discussion (always have in mind that you are teaching about science and not just science content)

- Always open and close your activities: Don't stop until you have developed a complete model of whatever phenomenon you are investigating or discussing.

As of this writing, I am not aware of any published instructional materials that incorporate MBST terminology. Your success depends upon your ability to frame and explain contextualized science to your students in your own words, using appropriate terms. Because of the power of language to shape our thoughts, regular thinking in MBST terms actually can change the way we design the science curriculum; which, of course, is the primary point of this book.

Summary

MBST is most consistent with, and builds upon, the principles of inquiry defined by the National Science Education Standards (NRC 1996). Implementation of MBST requires you as a teacher to plan and administer your science curriculum using a specific terminology—what we might call the "language of models"—at a level that is appropriate for the age and maturity of your students. If your students can do inquiry, they can understand MBST ideas about the nature and context of science at their level. It is better to infuse ideas about human thinking and learning into the curriculum as part of the process of scientific model-building than it is to treat it as a separate topic.

As a rule, you will be most successful with MBST if your primary goal as a science teacher is to teach students how to construct and evaluate scientific models, both mental and expressed. Learning of content and processes will follow from this goal.

Practical work in the lab or field should always pose a problem, and that problem should be how to construct a reliable, valid model of a natural phenomenon. Students assuming the role of problem solvers and model-builders focus their attention on planning and constructing the best possible descriptive and/or explanatory model of the phenomena they study. Students so engaged will necessarily learn the desired conceptual content.

Even if you can only lecture, you can still infuse MBST terminology and explanation into your discourse by referring to scientific models, theoretical models, and the creation of research based models. Because most students, especially at the higher grade levels, have some reasonably accurate notion of what a model is, the constant and consistent use of the term *models* can be expected to carry over to their understanding of the content you are presenting. While lecture-only science is not recommended, it does necessarily preclude the understanding that comes from using MBST.

In the next chapter, we will continue developing ideas about how to implement MBST in the classroom.

For Discussion and Practice

1. Analyze the idea that MBST principles regarding the nature of science and human thought should be commensurate with students' ability to grasp principles of inquiry. What does this mean to you in terms of students you have worked with or are preparing to work with?

2. How does the idea of *building a model* of a concept such as density differ from the notion of *teaching the concept* of density? Does the idea of modeling change your approach to planning in any way? How?

3. Some teachers might be concerned about introducing their students to the uncertainties of science. Review some of these uncertainties (the uncertainties of models and limitations of model building in general) and discuss how they might best be addressed during discussions with your students.

4. Our goal in MBST is to engage our students as model builders. How can you do this consistently, so that building a good model becomes the focus of their efforts?

5. A rather simple science activity, often used in teaching physical science, involves shining white light through a slit, and then through a prism to project a full spectrum. Filters are then used to change the color of the light, and students note changes in the projected spectrum. From this they can infer that the filters are absorbing certain colors (wavelengths) and allowing others to pass through. Discuss how you would frame this activity to make it a model-building activity. What would you tell students? How would you introduce the activity and pose the problem using MBST terminology?

References

Covey, S. 1989. *The 7 habits of highly effective people.* New York, NY: Fireside.

National Academy of Sciences. 2005. *America's lab report: Investigations in high school science.* Washington, DC: National Academies Press.

National Research Council (NRC). 1996. *National science education standards.* Washington, DC: National Academies Press.

Novak, J. D., and B. D. Gowin. 1984. *Learning how to learn.* New York, NY: Cambridge University Press.

Chapter 5

BUILDING MODELS in the CLASSROOM

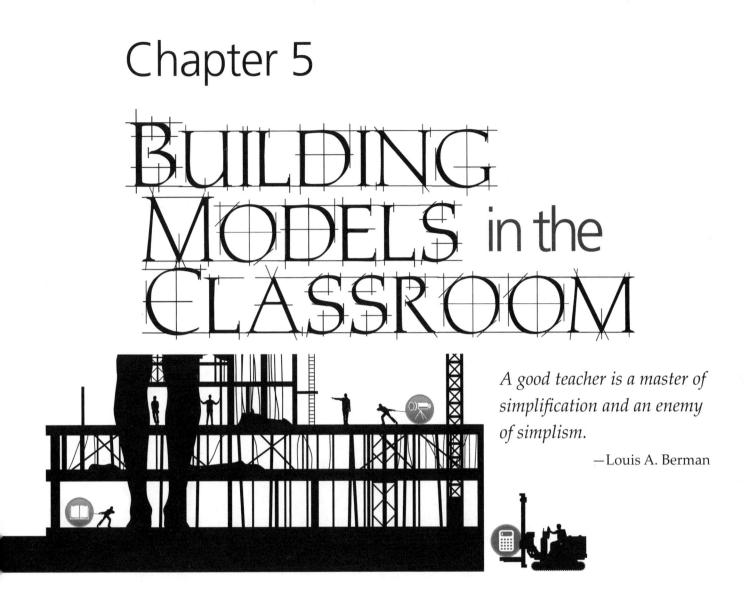

A good teacher is a master of simplification and an enemy of simplism.

—Louis A. Berman

If you've ever painted a house or any large object, you may have discovered a peculiar thing: that a light touch is not only easier, but is more effective than a heavy one. By allowing the brush to do the work, you save your wrist and the job comes out better. In a similar way, a washcloth normally gets you cleaner with a lighter touch. When you mash down too hard, the nubs of the rag cannot do their work. Good golfers know that their shots will be better if they let their club do the work.

The same principle is true in teaching. The master teacher plans carefully and then lets the plan work. The activities she creates are her instruments. If plans are well conceived and purposeful, they should achieve their desired effect.

Some educators suggest that lab and field (a.k.a. practical) work is a waste of time. That, of course, depends upon your goals. Practical work will be a waste of time when students and teacher do not have clear reasons for doing what they are doing; that is, when the outcome doesn't add value to a student's education. Unfortunately, the National Academy of Science's Board on Science Education (2005) has found considerable uncertainties among high school science teachers about the purpose of labs

Reaffirming subject matter content already taught and learned by students is probably not a valuable use for practical work. Nor is the motivational factor alone a very convincing argument for it (students tend to like lab work).

Elementary- and middle-level teachers may feel more certain about the purposes of hands-on work, but some authors have reported mixed results in terms of student learning outcomes.

In the last chapter, we laid out a specific purpose for lab or field activity: to teach students the fundamentals of constructing valid and reliable scientific (or at least quasi-scientific) models. When students are engaged in such learning, it was argued, content, process skills, and motivation take care of themselves. Not only that, but students gain more understanding of the goals of science.

Just as you don't paint better by slapping harder with the paintbrush, you don't teach better by slapping the students with more content. Students must learn within a framework that provides them with understanding and context. This principle of teaching is often expressed as *teaching less and teaching it better*.

In this chapter, we will go into more detail about using MBST in the lab and field, focusing on the practical concerns and techniques of model building in the science classroom. I won't apologize if I repeat some of what I wrote in the last chapter. Think of it as a spiral curriculum, coming back again and again to the same principles in different contexts. My hope, after all, is that you will become familiar enough with these ideas as to feel comfortable using them. We will necessarily be brief, and we will center our discussion upon the models within models commonly found in science. My hope is to give you a new perspective on common and familiar systems.

Parsimony and the Organization of Instruction

It is a well-known principle of thermodynamics that each time energy changes form, some of the energy is lost. In information theory, a similar principle holds, except that when information changes mediums, some information is often transformed in meaning, either gained or lost, or both.

A novelist writing a book is transforming the mental model in her head into a physical model consisting of words, syntax, symbols, and spaces. This model does not contain the full essence of the story as conceived by the author because the author cannot express fully what she means, sees, hears, and feels in her mind. All she can do is transfer symbols to a page. Information is lost as some is transformed.

BUILDING MODELS in the CLASSROOM

When we read the novel, we must construct the story from our own experiences, which are different from that of the novelist. We may gain some of the essence of what she intended, but the story really becomes ours. Information is lost again, but some is gained as well: but it is different information. The story I experience is not the same as the story the author experiences, nor of any other reader. Any teacher who believes that he can transfer models to students with the same understanding that he has for them is sadly deluded. It can't be done. Every student understands science in his or her own unique way—one that will meet his or her own needs.

Whether the model is created by a novelist or an architect, an artist, scientist, or teacher, model building is a creative act that involves making judgments—based on experience—about the purpose and composition of the model (see Table 5.1). Through our expressed models, we transfer the information to receivers who use it to reconstruct the model in their own minds.

Significant errors are less likely to occur when the model we are communicating is relatively simple and is consistent with models the recipient already has. More facts do not necessarily help learners strengthen their mental models. Too many facts can actually make a model cumbersome.

TABLE 5.1

Differing Models by Purpose

Function	Model Origin and Purpose
Novelist	Creates a written model of an internally perceived alternative reality with plausible elements reorganized imaginatively to entertain and possibly to instruct a reader. Reader must reconstruct the model mentally so that he/she experiences the story as a quasi-reality
Architect	Creates a model in several media depicting her internal model of a physical facility, a building, so that others can mentally reconstruct and assemble the elements needed for the completed structure
Artist	Creates a model in an artistic medium representing an interpretation of a perceived internal or external target according to personal internal standards and/or the external standards expected by a patron or employer
Scientist	Creates a predictive, data-based model intended to describe or explain some aspect of objective reality such that the model could be replicated by others using the same procedures

New teachers in particular often teach too much content without knowing how to develop the mental linkages students have to have to process the new models. In fact, this is one reason why expertise alone is not sufficient to make one a good teacher. Some experts lack the ability to tease out and organize the core elements of the models they present to their students. They present information. They do not build models.

Mental model building only occurs when the mind (not necessarily your conscious mind) sees a purpose for it. You will find that your worst students are often bright but lack a sense of purpose. What purpose, they ask, do the concepts of force, osmosis, chemical bonding, or plate tectonics have for me?

You must decide what kind of model will serve them best. A mind willing to accept a general model of a phenomenon may resist spending time and energy adding details to a model that seems unnecessary and confusing.[1] In general, you are more apt to succeed if you assemble conceptual models inductively, moving from simple facts to simple principles, then moving deductively to create a firmer principled model with additional facts.

The best teachers will begin with the end in mind. They provide a framework. When I was in college, I tried to learn calculus without knowing its purpose. The instructor already knew and didn't feel any need to develop such a foundation. Without that meaning, my efforts foundered. The same principle applies to any area of teaching. If I'm an English major, I might take it for granted that I should learn Shakespeare, but my efforts will mean more if I had well-defined goals consistent with my life goals.

In science, as in most fields of study, models are interrelated. Students must understand individual models before they can understand composite ones. So, for example, students must understand mass and acceleration before they can understand force, since force is by definition a relationship between mass and acceleration ($f = ma$, or force equals mass times acceleration).

This means a good teacher must do two things: simplify the model and ensure it is understood. Some teachers may interpret this as a call to "dumb down" the curriculum but simplification is *not* the same as simplism. Complex scientific models are constructed one relationship at a time.

You cannot beat good planning when it comes to the technical component of effective teaching. Know the facts, then know how the facts come together to create a viable model. Plan well and let your plan do the work. Your strategy should be to engage your students in constructing the simplest, most lucid model of a phenomenon that will serve your ends. Then enrich the model, avoiding the temptation to present too much information too fast. Make sure students understand a model before moving on.

In teaching, unnecessary jargon complicates our models. Experts often make models unnecessarily complex. Avoid throwing terms at students just because "they should be exposed to them." Terms without a firm grounding are like free radicals in biological systems: They run around doing damage to established models. Avoid them.

---

[1] Time and energy are both biologically important adaptive factors in humans but wasting either is generally maladaptive and avoided by thinking animals—especially wasting energy. Confused mental models can be especially risky in nature. Students often resist models that confuse their existing models.

BUILDING MODELS in the CLASSROOM

The Spiral Curriculum

Many teachers organize their instruction in a linear sequence, often following the sequence of topics in a textbook or a set of school district curriculum guidelines. You may be externally constrained in your planning, but if you have enough freedom you should consider avoiding this pattern. Linear, sequential planning often results in a rush to "cover" a topic, which leads to inadequate development of the model. Alternatively, you may end up dispensing with important models that fall toward the end of the school year.

An alternative model is to create a *spiral curriculum* by introducing all of the core conceptual models early in the year and then developing them over the course of the school year. This plan is based on a principle familiar to anyone acquitted with journalism: that of *front loading*. Journalists are taught to "front load" information so that if it becomes necessary to cut a story, an editor can simply lop off copy at the bottom without losing the essential elements of the story. Educators adapt this principle to teaching by revisiting concepts repeatedly, building simple models up by enriching them and relating them to one another.

A highly simplified example of how a conceptual model is enriched by addition over the course of the school year is shown in Table 5.2, using the gene concept as its focus. The spiral curriculum model requires you to know the most important concepts—the core concepts—around which you are developing your curriculum. Whether you are teaching general science, chemistry, biology, or one of the other sciences, you should be able to identify these concepts.

Here's a task. Set aside this book for a moment and make a list of the 18–24 major conceptual models that you believe encompass the scientific content that you teach

TABLE 5.2

Additive Development of the Gene Concept Over a School Year

Quarter	Exemplary Propositions
First	Genes are segments of hereditary material that are the blueprints for protein
Second	Genes are encoded by bases in DNA and transcribed in ribosomes: Proteins determine structure and function of cells that form single cell and multicellular organisms (relationship to cell theory)
Third	Mutations in genes lead to variations in cellular function and/or appearance; organisms with variations may be selected by agents in the environment leading to evolution of diverse life forms and formation of new communities (relationship to cell theory, evolution, ecosystems)
Fourth	Genes can be transposed naturally by recombination; artificial recombinant technology can allow humans to bioengineer altered life forms (relationship to technology and human welfare)

your students. Your students absolutely must understand these conceptual models to be scientifically literate. What conceptual models would you choose if these were the *only* models that they would learn all year? These core mental models are the organizing models you should use—the backbone of the composite content model you intend to present to your students. Metaphorically, they will be the linkage points in the web of relationships you plan to build. If you still need more metaphors, think of a building with exposed uprights and roof beams upon which the whole structure depends. At the top of the rooms is one long beam to which the lateral beams attach and upon which they depend. In our mental models, the most important "beams" are our core models, organized into an interdependent hierarchy.

Once you have identified your core mental models (key or core concepts), you will find that some models are more important than others are. That is, they seem to explain a great deal. You should think of them as *thematic models*. Thematic models are mental constructs that seem to anchor and unify the core concepts. In biology and the Earth sciences, *evolution* is a major thematic model. In physics, *force* appears again and again in various models; while in all of the sciences, *energy* and *entropy/negentropy* are thematic models. Themes give backbone to your overall model and thus strengthen its framework.

Your next step in planning is to envision how you will introduce all of these models early in the year and then enrich them as you tie them together over the course of the school year. One way to do this is to incorporate thematic models as developmental organizers as shown in Figure 5.1.

The spiral curriculum model can be implemented at any grade level. An essential feature of the spiral curriculum is the constant modification and enrichment of the same core models: a process that strengthens the structural function of these models in your students' mental models.

Your total curriculum, of course, will include supplemental models that support the core and are important in their own right, but your focus is on building the structure—adding models only as you need them to complete the job.

As a teacher, you are a builder. You build by placing and attaching one model at a time. Since it is impossible to construct a complete model (too many parts), the models you use to support your core models are exemplars. An *exemplar* is a model that we mentally reference as an archetype for a whole class of phenomenon. The lily, for example, is often used by biology teachers as an archetype for all flowers. The acid-base reaction lab in chemistry represents all (or most) acid-base reactions in some essential way. The pendulum in the lab is the archetype for all pendulums. You get the picture.

Good exemplars are easily understood, memorable, and broadly representative. They are easy to enrich. The most effective exemplar can represent a number of targets with only minor modifications. Too much specific detail diminishes the effectiveness of an exemplar because details differ from target to target. It is best to begin the year with simple exemplars and enrich them as you return to them.

Exploring and Explaining Analogies

Any model that is more than a simple symbol is analogical. Signs and symbols represent targets, but do not tell us anything about them. A stop sign on a street symbolizes a desired behavior but has no attributes corresponding to attributes of the desired behavior. Similarly, if I yell "stop" at you, the word symbolizes an action, but does not represent it analogically.

Used in a sentence, words become analogical in that each word in the sentence (and its position in the sentence) denotes a specific object, action, or condition. For example, each element of the sentence "Jack and Jill ran up the hill" has a corresponding element in our mental model representing the objects and actions. The phrase "the hill ran up Jack and Jill" does not have the same meaning, but does have the same words. A sentence is a syntactical model.

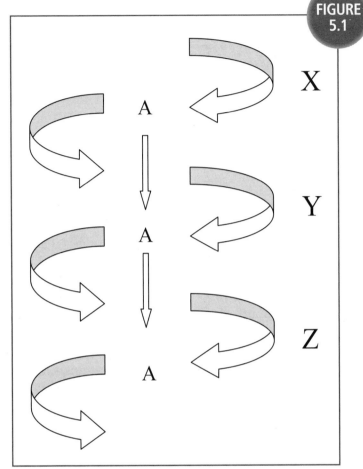

FIGURE 5.1

The spiral curriculum repeatedly returns to the same model (A) and enriches it by linking to core models (X–Z) over the course of the school year.

The label "2" has a symbolic meaning. The proposition "2 + 2 = 4" has analogical meaning that is not the same as "2 + 4 = 2." It is also a syntactical model.

My point is that meaningful models—those that allow us to understand and explain our subjective perceptions of the objective world—are analogical, and to understand any model, we must understand the relevant correspondences between the model and its target. We should not assume these correspondences are recognized, but rather we should make sure our students can accurately dissect a model and reassemble it. One way to do this is to ask them to explain or to visualize the objects, quantities, conditions, or relationships each feature of their model represents. Take, as an example, the law of gravitation, expressed as

$$F = G\left(\frac{m_1 m_2}{r^2}\right)$$

where

- F is the magnitude of the gravitational force between the two point masses,

- G is the gravitational constant,

- m_1, m_2 are the masses of two bodies, and

- r is the distance between the centers of the masses.

When students have problems in math and science with such models as this, it is formally because they have not *imagined* what each symbol represents. By imagining, I am referring to visualizing a mental image that gives the symbols concrete meaning. For example, set two balls, a basketball and tennis ball, about a foot apart. These are two bodies with mass. Ask, "What symbol in the equation represents their mass, and what represents the distance between them?"

Now visualize G as a constant force pulling these masses toward each other. The force is far too weak to cause them to move, but it is there. Now move the balls two feet apart. What has changed in the equation? The distance between them (r) has doubled to two, meaning r^2, which was 1 (because $1 \times 1 = 1$) now becomes 2×2, or 4. The force (F) is now ¼ of what it was because the masses are the same and G is a constant.

Visualize the force as a white ghost cloud between the balls. As you move them apart, the cloud gets thinner. How much thinner? If you double the distance, the cloud thins to ¼ of what it was. If you moved the balls three feet apart, the cloud would only be ⅑ what it was ($r = 3$, and $r^2 = 9$).

Chemistry and physics are often difficult for students because mathematical models such as these are never imagined as real entities. Visualization coupled with explanation can be a powerful tool in learning and understanding. Explaining the model requires visualizing and explaining the target system and you cannot say you have truly learned something until you can do this. You should require your students to explain every new significant model they encounter, as shown in Figure 5.2.

Models in School Science Reports

The goal in creating a research report, whether a simple one-paragraph summary or a formal report, is to establish clear analogical links between the system under study—the target—and elements of the model. The clearer these linkages are, the better the model.

When a scientist constructs a report, she carefully selects the components to include in the model. The model includes a clear representation of what she did, what she observed, and what she thinks her results represent in the immediate system she has studied, in other related systems, and in the larger explanatory (theoretical) model.

FIGURE
5.2

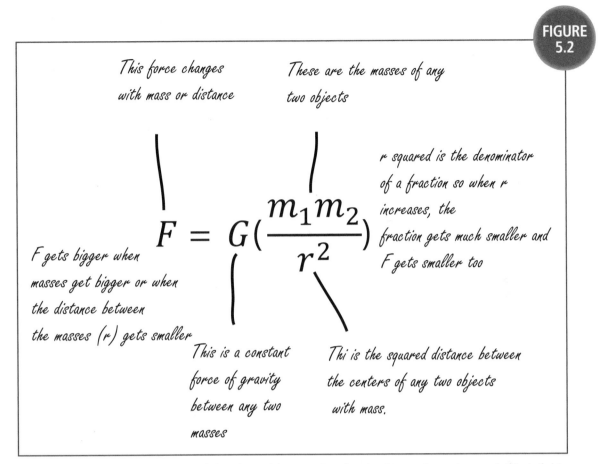

Asking students to explain a model such as this equation for the force of gravity can be helpful in ensuring that they know and understand what each component of the model represents.

For the most part, scientists have an idea of what the model will look like when they begin. Each model is a project that requires planning and forethought. Scientists engage in open exploration and discovery, but they have to construct meaning from what they find, which is the true essence of science—not the exploration and discovery itself. They create knowledge from the facts they find.

Our ultimate purpose in school science is to teach students how to construct valid and reliable models that clearly express the mental models of the natural phenomena they study.

Let's look at the characteristics of some of the models your students might construct as they describe and interpret their findings in the lab and field. These models—their expressed models—are comprised of verbal, mathematical, diagrammatic, pictorial, symbolic, and, occasionally, three-dimensional elements.

Students will not learn how to assemble good models on their own without guidance. Because MBST requires multiple data sources in order to assess reliability (even if only to estimate it), most labs are best done as a joint-class project, with individuals or teams pooling their data for analysis. The class as a whole should also participate in selecting the elements of the model, since the process of selecting such elements is a key learning experience. Simplicity and clarity are the standards for good model making.

Verbal Modeling

Verbal explanations are the heart of all final models. You cannot present a meaning-ful complete model without verbal statements. Verbal models in research reports most often take the form of

- Problem statements: *What did we want to do?*

- Theoretical analyses: *Why do it?*

- Hypothetical models, including predictions: *What did we expect to happen?* (if a test)

- Concisely written directions and procedures: *What did we do?*

- Data analysis and summary: *What did we find?*

- Discussion and conclusion: *Why is it important?*

Problem Statements

Problem statements concisely define the problem the *students* are solving. Published lab activities and those in lab books often do not have good problem statements. In science, the problem is not to explore, learn, illustrate, demonstrate, or discover something. It is to build a good descriptive or explanatory model of it.

Therefore, in MBST it is recommended that you frame the problem in the con-text of model building. In your lesson plans, you might express the problem as, "the students will build a [descriptive or explanatory] model of [whatever phe-nomenon they are studying]."

Students may use similar language, but from their perspective. Elementary students might write that they are building a model of how gravity acts on objects of different weights. In high school, students might be somewhat more formal in their recognition of the kind of model they are building: "The problem in this study is to build a descriptive model of influence of gravity on objects of different masses and similar size and to suggest explanations for any variations we might see."

If your biology students are studying cells for the first time, their problem may be to create a descriptive model of a cell, or to compare the form and structure of cells with their different functions and suggest reasons for the differences. Such a problem is more representative of the goals and purpose of science than to "dis-cover" the structure of a cell.

BUILDING MODELS in the CLASSROOM

Table 5.3 provides examples of problem statements in MBST format for comparison to actual problem statements taken from real activities.

Theoretical Analyses

In scientific models, scientists usually explain how their model adds to existing theoretical models. In MBST, you can approximate this by engaging your students from the first day in the development of a personal theoretical model that represents what they have learned in the class. Since your purpose is to build forward constantly, you should continuously be relating any new models students are building to models they have already created.

The composite mental structure of the models they have already created represents your students' theoretical model. By continually enriching their growing model, you create a sense of progression, growth, and meaning. No model should be created in isolation from others except at the very beginning, and event these early models should be related to some meaning aspects of the students' lives. If your students cannot relate their new model to other models they have completed (or at least to some aspect of their lives), the new model will have little purpose.

TABLE 5.3

Problem Statements in MBST Format

Problem Statement	MBST Problem Statement	Comment
Find out how water bends (refracts) light that enters it.	Create a model of the behavior of light as it enters water from various angles.	The problem statements from actual labs are stated as objectives rather than problems. In fact, none of the labs had real problem statements. The labs themselves may have been traditional or inquiry in format but none of them really reflect the purpose of science. The MBST statements, on the other hand, might better reflect what a scientist would set out to do.
Investigate the effects of changes in mass and slope on a car's acceleration.	Create a mathematical model of the relationship of mass and slope to a car's acceleration.	
Light a bulb by creating a simple electrical circuit.	Create a descriptive model of a simple electrical circuit.	
Investigate the cellular basis of variation in organisms by exploring the processes involved in sexual reproduction.	Create a model of the observable changes in DNA in cells involved in sexual reproduction .	

Hypothetical Models

Hypothetical models are speculations on *why* a relationship exists or a particular effect is seen. They are *explanations*. This meaning of hypothesis is often confused with simple *predictions*. You are not testing a hypothesis if you just make prediction without a clear reason.

A hypothetical model is only necessary when you are testing an explanation. Hypothetical models can be stated in several forms:

- As a speculative *explanation* for a phenomenon: "I think the reason the system is running so slowly is that it needs oil."

- As a more formal *explanatory* statement in the form "if A, then B," where A is the hypothesis and B is the consequent: "If friction in the system is high due to the lack of oil, then it will run more slowly than it should."

- As a *research hypothesis* that takes the form of a prediction *based on our proposed explanation*: "If friction in the system is high due to the lack of oil then reducing the friction by oiling the system will make it run significantly faster."

Hypothetical models are generated whenever you ask for an explanation of something unknown. If you ask fourth graders *what* they think steam is and *why* they think steam comes from water when it gets hot, you will receive hypothetical responses (if they don't already know). If your students think that the heat is causing the steam, they may be able to suggest (with some guidance from you as a teacher) that cooling the steam will cause liquid water to reappear (a logical prediction). They can test their hypothetical model that steam is just very warm water by holding a glass filled with ice in the steam column and observing water condense on the glass.

You might then see if something similar happens with liquids other than water, using the same procedure. Carefully warmed rubbing alcohol[2] and hot cooking oils could be tested in a similar way, thus you could build a rich model of *vaporization* using similar liquids.

In the text box, you see a lesson plan for an activity on inertia commonly used in elementary and middle level science that includes hypotheses. Notice that they become important only after the initial discovery, to test an explanation. Asking for predictions otherwise (in the introduction to a discovery, for example) may raise interest, but it is not hypothesizing unless it is tied to a hypothetical explanatory model.

Qualitative Descriptive Models

Qualitative description plays an important role in science. Although natural scientists tend to favor numerical data due to its precision, numerical data are always presented in association with a qualitative description of them.

As data, qualitative description is less precise than quantitative data and often more open to interpretation. The apple that I call red, you may call pink. Precise description is important in scientific modeling. The ambiguity associated with qualitative data is one reason why scientists prefer to work with numbers.

One way to measure qualitative data more precisely is to convert it to numbers via a scale or rubric that assigns qualities to numbers. Another way is to create

---

2 Any alcohol is flammable. Never warm it with an open flame. Gentle warming of volatile isopropyl alcohol by setting it in a warm location may be enough to create desired amounts of vapor.

Concept: Inertia

Needed: A raw egg and a hard-boiled egg

Tell the Students: I have discovered the most amazing thing! I was hardboiling eggs when I got them mixed up with raw eggs. As I sat there trying to figure out how to tell which was which, I spun them idly. Then I noticed something peculiar when I stopped the eggs from spinning. Diane, would you help me by spinning this egg?

Discovery: Spin both eggs. The raw one will resume spinning when you stop it and let it go. The hard-boiled egg will remain stopped. Crack the eggs open to see which egg is which.

What You Know: The raw egg will continue to spin because the liquid inside has not stopped moving. Its inertia, the tendency to continue to move unless completely stopped, will cause it to start spinning again. The spin must be hard enough to allow this to happen.

Building the Model: Tell your students they are going to construct an explanatory model. Have them describe exactly what happened and to suggest why. Tell them to try to visualize (imagine) what's going on inside each egg and what the difference is. Remind them that differences in variables can explain differences in behavior. Both have yolks; both have whites; one is solid and the other is liquid. How could the difference explain what they saw? If no one suggests that the liquid is moving, tell them to imagine water in a bucket if the bucket is spun around and then stopped. At this point, someone is likely to suggest the desired initial hypothesis below. Notice the steps in building a test:

Preliminary explanation (initial hypothetical model): *The continued movement of the liquid causes the egg to move.*

Formal hypothetical model: *If the liquid continues to move, then the spinning egg will resume moving after it has been stopped.*

Prediction: Therefore, *if the liquid is turned solid by boiling the egg, the egg will no longer move once we stop it.*

Have your students complete the test and their model. Apply the term *inertia* to the model to describe the continuing movement of the liquid in the egg once it is spun and stopped. You can enrich the model by spinning a bucket of water on a potter's wheel and then suddenly stopping the wheel, observing what happens to the water. Relate this to the egg.

categories based on certain qualities and count the number of items in each category (categorical data).

The purpose of qualitative data is generally to reveal patterns that may suggest relationships and explanations. Qualitative models are found more commonly in newer sciences, on the frontiers of science, and in sciences where specific numbers may be misleading. In the case of the newer sciences and the frontiers of science, scientists tend to be more descriptive and inductive, looking for patterns that might lead to puzzles, explanations, and testing.

Numbers may be misleading when models contain variables that are inherently unstable and unpredictable. The less predictable and changeable the phenomenon, the more one must focus on broad patterns rather than overly specific numerical data.

Clear reporting is the key to good qualitative data. In the natural sciences (as opposed to some social sciences), most scientists try to keep themselves from pondering the meaning of data while they collect it, for fear that it will bias their perceptions.

Procedural and Data Models

A verbal model always accompanies data tables, charts, pictorials, or any other forms of data-based models. The procedural model describes how the model was built precisely enough so that colleagues with the skills could replicate the model if they wished. Peer reviewers are very interested in knowing how the data were collected, just as juries, judges, and attorneys are interested in knowing how evidence was collected prior to a trial.

For the most part, raw (unprocessed) data are not presented in a research model unless the numbers are so small that averaging would be unreasonable. What readers are interested in is the substantive data analysis that demonstrated that the data support the model's conclusions.

The data, like items of evidence in a criminal trial, will make or break the case. You should not present more than is necessary but, alternatively, you should include enough to be convincing. Keep in mind that a research report is an argument, but also keep in mind that one of the tenets of science is make the model fair and aboveboard.

Models of Results, Discussions, and Conclusions

Scientific models include results, discussions, and conclusions, although they may not be divided into separate sections in the model. The *results* summarize findings that bear on the problem addressed by the model. They draw attention to findings in the data, but do not state conclusions about those findings. Concerns or weaknesses in the data may be stated here as well. In a legal trial, the results section is comparable to a summary of the evidence.

The *discussion* is an active intellectual examination of the results in relation to the stated purpose of the model. In this section, the scientist knits the data together through analysis and synthesis. If hypotheses are being tested, he shows how the data support (or do not support) them. In a courtroom, this would be comparable to the middle of a lawyer's summation, in which she constructs a narrative from the summary of evidence. The key word used to describe the discussion section is *interpretation.*

Building Models in the Classroom

The *conclusion*, in a courtroom, is a statement of the major inferences that arise from the evidence and its interpretation. In a scientific model, the conclusion summarizes the *logical inferences* one should draw based on the findings, and relates them to existing theoretical models. Conclusions in both science and law often include a summary of the *benefits* to be gained by accepting the model. In science, this might include potentially fruitful directions for new research.

This three-part structure characterizes most models for argument, in science or in other fields. A scientific model is largely argumentative, *argument* in this case referring to reasoning aimed at achieving acceptance. Hypothesis testing is clearly argumentative by nature, but even purely descriptive models will be argued, since they must be accepted by the larger scientific community.

Mathematical Models

Scientists value mathematical models. Historically, science began as an effort to quantify experience. It has matured since then, but mathematics is still highly valued. Mathematical models add value to a model only when the data collected are reliable and valid, conditions we discussed in a previous chapter. In this section, we will focus on data summaries in the several forms commonly found in professional and school science including

- tabular data models (data tables),
- statistical models,
- equations,
- formulas, and
- graphical models.

All of these models have a syntactical structure, meaning the position of the symbols is as important as the meanings of the symbols themselves. All function through analogy, but none of them is transparent on its face; they all require the application of rules of interpretation.

Tabular Data Models

A tabular data model, or data table, is one of the most common models found in research papers. A data table presents data in rows and columns, following a formally accepted pattern. I will take it for granted that you are familiar with models such as Figure 5.3, which compares the rates of flow of two different liquids at three different temperatures. Tables are often used for comparison.

The important thing to emphasize to your students is that each element of a table has a real correspondence to some element of the target. It is a good idea to work with students to ensure that they are able to map models to targets. Visualizing what each piece of the model represents can be especially helpful.

The rules for creating tabular models are reasonably consistent in various guides. Most guides recommend they be numbered, titled above with a short title making the purpose of the table clear, clearly labeled, and include clearly labeled units of measurement.

Statistical Models

Statistical models are rare in K–12 and will not be discussed much here. When they are used, they formally have a tabular format that you must use in the model. Computers are normally used to generate statistics now, and they generally come programmed to create the model in the proper format.

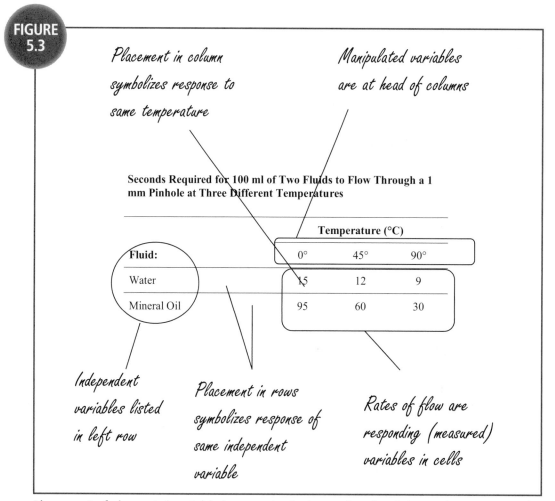

FIGURE 5.3

Placement in column symbolizes response to same temperature

Manipulated variables are at head of columns

Seconds Required for 100 ml of Two Fluids to Flow Through a 1 mm Pinhole at Three Different Temperatures

Fluid:	Temperature (°C)		
	0°	45°	90°
Water	15	12	9
Mineral Oil	95	60	30

Independent variables listed in left row

Placement in rows symbolizes response of same independent variable

Rates of flow are responding (measured) variables in cells

Placement of elements in a table have syntactical meaning that is just as important as the meanings of the various symbols. Each element corresponds to something in the real system that is the target of the model.

Formulas and Equations

Formulas and equations are symbolic and syntactical models. As we discussed earlier in this chapter, it is very important for students to understand both the symbols and the syntax for each formula.

Because you may already be familiar with these models and because you feel the pressure of time, you may be tempted simply to give your students a formula or equation and assume they will understand it as they use it. This leads to a great deal of rote learning, as students plug numbers into a model they don't really understand. To avoid this, take the time to make sure that your students fully understand the elements (symbols and placements) of the models they use. The remedy for this was suggested earlier:

- Ask students to precisely explain the model, translating each part of it into words and asking questions about any element of the model that they do not understand.

- Have students mentally visualize the elements of the model and, if possible, create a concrete representation of all or part of the model.

- Work with the model concretely, as when we moved the balls apart in the gravity example.

Graphical Models

Scientists and science teachers commonly use graphs (graphical models) to reveal or portray significant patterns in data. These are a kind of pictorial model, but because of their mathematical nature, we will include them here. The most common models in K–12 are the stem and leaf models, histograms, line graphs, and pie charts (Figure. 5.4). Your choice of model depends upon whether the data collected are categorical or continuous. Although graphical models vary in their constructions, most have two reference lines: a horizontal (x) axis and a vertical (y) axis. By convention, most models place values of the manipulated (independent) variable on the x-axis and values of the measured, counted, or responding (dependent) variable on the y-axis.

Categorical data are collected by counting the number of members in a category or collection of like things: i.e., how many people have brown hair; how many robins are eating on the lawn, and so forth. Categorical data are often represented on a histogram or bar graph.[3] In both models, the length of each bar represents either the absolute number of events or the frequency (in percent of the whole group) in each category. Stem-and-leaf plots (Figure 5.5) are a variation in which individual

---

[3] A histogram is a bar graph whose bars represent categories that can be arranged in a progression, as from smallest to largest in five groupings, for example, or days of the week from Monday to Sunday. In bar graphs that are not histograms, the categories are not progressive: comparing numbers of apples, oranges, and grapefruits, for example.

events are marked with "Xs." Bar graphs are sometimes turned on their sides but the principle behind them is the same, regardless of their orientations.

In contrast, line graphs and scattergrams are used to display continuous data. *Continuous data* represent variables for which we assume there are an infinite number of values, any one of which we could conceivably measure.

For example, let's suppose you heat a beaker of water from 50°C to 95°C. If you created a graphical model of this change, you would end up with a line graph

FIGURE 5.4

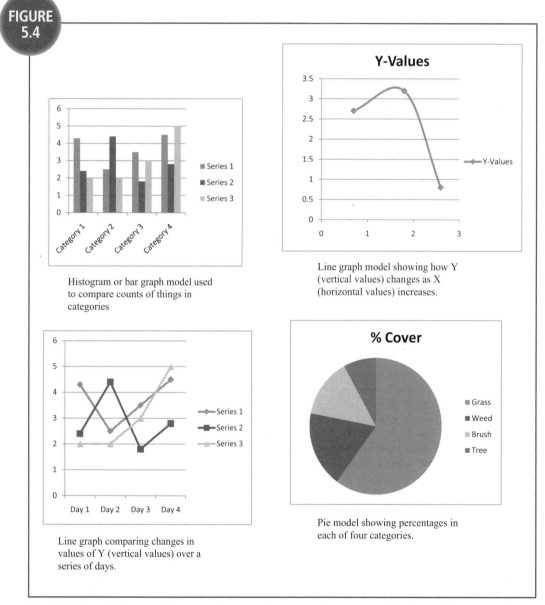

Histogram or bar graph model used to compare counts of things in categories

Line graph model showing how Y (vertical values) changes as X (horizontal values) increases.

Line graph comparing changes in values of Y (vertical values) over a series of days.

Pie model showing percentages in each of four categories.

Graphical models come in many forms and are used to present both continuous and categorical data.

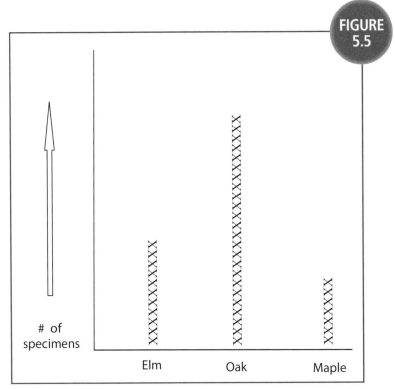

Stem and leaf diagrams such as this are simple ways to display categorical data. The graph may be flipped so the bars are displayed horizontally.

like that in Figure 5.6. Along the line from point A to point B are an infinite number of temperatures and times. You may not measure them, but they are there, nonetheless.

Even if you measured temperature every 10 minutes and used only those temperatures on your graph, it would be inappropriate to build a line graph model. The nature of time and temperature is that they are continuous variables and a line graph is appropriate.

But, you say, what if I were to measure the number of events that occurred in five-minute intervals? Now you are counting rather than measuring at least one variable. Each five-minute interval would be a category, so a bar graph would be appropriate, not a line graph.

The difference between how we display continuous and categorical data is illustrated in Figure 5.7. When we want to show a continuous change in a variable as it moves though a range of values, we usually use a line graph. When we want to compare amounts or frequencies in groups, we use a bar graph.

Diagrammatic and Pictorial Models

Diagrammatic models, or *diagrams*, are two-dimensional models symbolically representing objects and relationships according to some set of conventions. This category of models includes, but is not limited to

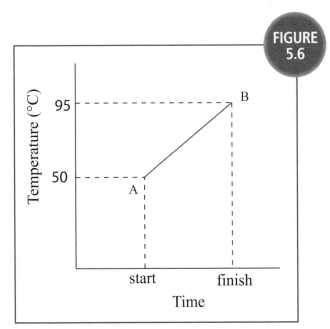

Graph of water heated from 50°C to 95°C.

- flow charts

- Venn diagrams

- blueprints and schematics

- outlines and concept maps

- fishbone and stem-and-leaf diagrams

- maps and plots

Diagrammatic models represent objects or processes figuratively (metaphorically): a double line on a blueprint stands for a wall, for example. A relationship is an arrow. As with mathematical models, diagrammatic models usually require rules of interpretation to understand them.

Pictorial models, in contrast, represent the literal (actual) appearance of objects and we usually link them intuitively to their targets. They include photographs, drawings, some computer simulations, cartoons, and so forth.

We create both types of expressed model from, or in response to, the holistic visual imagery contained within our mental models.[4] Although diagrams do not represent the literal appearance of objects we perceive, our brain processes them as pictorial models. According to Bertel (2005),

> Diagrams have been related to visual mental images, either by computational or representational metaphors. Cognitive mechanisms involved in the inspection of diagrams and those involved in the construction and inspection of mental images are found to interface at later… and earlier stages of mental processing….

Diagrammatic models are educationally useful because (a) they allow students to visualize relationships holistically and (b) most diagrams require students to engage with them as they interpret them. Unfortunately, many teachers do not recognize the educational value of these models, treating them as supplements to, rather than integral parts of, mental model building.

You can employ diagrams and pictorials in a number of ways: by having your students make them, analyze them, and create associative networks among them—all activities that should yield benefits in terms of mental modeling. The sages of journalism had a reason for saying a picture is worth a thousand words.

The picture of the dog in Figure 5.8 could easily anchor a conceptual model of carnivores. You can learn a great deal about adaptation in carnivores just by studying the features of this animal: the perked-up ears, forward-looking eyes, large nose, tooth and mouth shape, and leg/body structure all are telling clues about its natural lifestyle (or at least the lifestyle of its forebears). But if we just think of it

---

[4] Imagery corresponding to visual imagery also exists for our other senses such as auditory or tactile imagery, but is not usually relevant to school science unless we are dealing with other media. You may argue that the photograph is not a model we create because it is not created in our mind first. We do create it with our technology, though, and give it meaning when we view it.

FIGURE
5.7

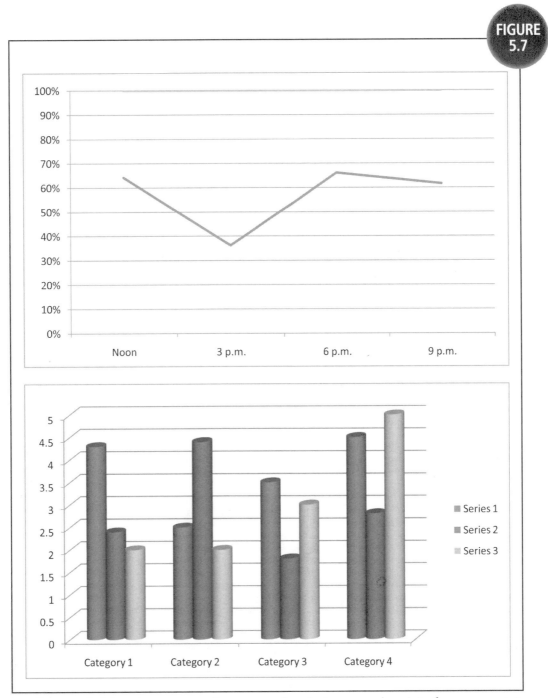

In a line graph (above) we assume there are values between each point of measurement, so at 4:30 p.m. the value of variable *y* would be around 50%. If you think of a histogram as representing stacks of counted items, you will see that there are no "in between" values. What you see is what you get.

as a decorative insert in a textbook or curriculum guide, we will miss its relevance and importance.

Diagrammatic and pictorial models are fitting subjects for inquiry. You can engage your students in meaningful true critical analysis, developing skills in observing and reasoning, by asking them interpret and explain these models.

How to Select Elements and Build and Evaluate a Scientific Model

By the time students leave high school, your students should be able to construct a scientific model of a phenomenon—a model appropriate for their level of knowledge. Why else should we use valuable instructional time to engage them in lab and field work?

If your students are to become successful model builders, they must do more than follow recipes that you provide for them. No one has ever become a master chef by religiously following other people's cookbooks.

Students without experience in scientific model building need guidance, of course. Younger children will be more adept at learning *what* than *why*.

In all cases, you should engage your students in planning the creation of their models rather than doing it for them. Just as parents who do everything for their children do them no favors, so the teacher who designs every learning experience cheats their students of opportunities.

FIGURE 5.8

This image model of a dog can be used to discover many features typical of amammalian carnivores, and so may be the basis fo a conceptual model of such creatures.

BUILDING MODELS in the CLASSROOM

Your primary job as a classroom leader is to present students with a solvable problem in terms—and at a level—that they can grasp. Discovery of a phenomenon is only the first step. You can usually set that up. The problem then becomes one of describing, explaining, or affirming a tentative explanation by building a model. In most cases, your planning of a model should be a whole-class activity rather than an individualized one. There are several reasons for taking this collective approach; among other things,

- students learn from one another;

- students "own" the investigation as a group;

- data can be pooled and analyzed, allowing for concurrent replication;

- errors become more apparent in pooled data;

- students can analyze and discuss the results together; and

- it makes your job easier while enhancing instructional quality.

The amount of guidance you will need to provide varies with the age and experience of your students. Young children typically need more guidance while older experienced students can contribute more to the building of the model.

Let's imagine you are a fourth-grade teacher interested in teaching your students about concave and convex lenses. To do this, you will need several kinds of lenses: single concave, double concave, single convex, double convex, and other variations including similar lenses of different power.

Following the MBST format, you show the students simple concave and convex lenses and let the students work with them. Before long they discover that the different shapes of glass can have strange effects on the appearance of the objects when you look through the glass at them. In the subsequent discussion, you ask your students to describe the different effects.

This is a typical *introduction to inquiry*. It poses a problem that can now be exploited to create a descriptive model of lens behavior. Setting up the problem to create a model of the phenomenon in a systematic way might require a discussion like this one:

T: *Scientists create models that describe and explain things. We might not be able to explain what's going on yet, but we can create a model to describe what we see. Our model should be good enough so that other people could look at it and understand what we saw. That's what scientists try to do when they create their models. We might even be able to suggest some reasons for what we see later on. So let's start by looking at what you found so far. Were there any differences in what you saw when you looked through the two different lenses?*

S1: *Uh-huh. Sometimes they were upside down. Sometimes what we saw looked bigger and sometimes it looked smaller.*

(The students describe the differences. The teacher suggests they call one kind of lens concave "because it looks like a cave" and the other kind of lens convex).

T: *What do you think might have caused these differences?*

S2: *Maybe the shapes of the glass?*

T: *So you think the shape of the lens might cause the differences in what we see?*

S2: *Yeah.*

T: *So what do we need to know for our model?*

S: *The shape of the lenses we test.*

T: *Okay, the lens shape is the variable we are changing—we are testing. And what changes when you change the shape of the lenses?*

S1: *How things look through the lenses.*

T: *So how things look depends on the shape of the lens, right? And since how things look changes depending on the shape of the lens, how things look is the dependent variable. That will be our data. How can we show how things look?*

S2: *We could draw pictures.*

S3: *We could describe what we see.*

T: *But if we all have pictures or describe different things, how can we pool and compare our data?*

S4: *We could look at the same thing.*

T: *So should we describe what we see or draw a picture of both?*

S1: *I think describing it will be okay. If we draw it, it would have to be something simple.*

S6: *Like the letter "R" on the wall there.*

T: *Okay, does everyone agree we will look at the letter R and describe what we see?*

(Students agree)

This decidedly idealized dialog illustrates roughly how you can involve students in planning the construction of an investigative scientific model of a phenomenon. Notice that the teacher keeps her students focused on building a model, leading them to consider each variable in turn that might be important in the final model.

BUILDING MODELS in the CLASSROOM

If necessary, she can dictate what will be in the model, but she allows the students some leeway in deciding what to include.

As students build models together and learn the jargon (*model, variable, data*) the process of model building should become easier and more efficient. Students will gain confidence and will take more responsibility for planning.

Planning as a group takes time, though, and some teachers may be unwilling or unable to make the time needed to plan well. Notice that the students are not so much engaged in planning what they will do as in how they will collect data and create the outcome model. Even if trade-offs are necessary for the sake of time, students should be involved in some aspect of the planning and they should always be conscious of the ultimate goal.

Finding and Accounting for Errors in Scientific Model Building

We spoke in earlier chapters of errors in models. Error is not the same thing as omission. An *error* is any deviation in accuracy or correctness, while an *omission* is simply the absence of some feature of the target. An omission may be deliberate or it may occur through error.

We think bad things of error, but it is unavoidable. Consider how many things you get wrong in your daily life. We make errors all the time because perfection is nearly impossible to attain. Here's the thing: error is not innately bad. It is a fact of life. Fortunately, most of our daily errors don't have serious consequences. If they are important, we try to correct them as best we can. If we can't correct them, we compensate for them by adapting to the new conditions the error creates. That's the way it is in life and that's how it is in science. It's doubtful that any scientific model is error-free, given the many sources of error innate in the model-building process. What are those sources of error?

In school science, as in professional science, *observer error* is a major problem. All observers, whether professionals or students, bring preconceptions into the lab or field that may influence their choice and recording of data. Our choices of what data to record and the methods we use to record them are subjective; we make judgments based on our expectations. Our perceptions of facts are also subject to error.

Even if we are just taking digital readings from electronic instruments, the instruments themselves are subject to error from miscalibration, damage, and the quality of their manufacture. This is *equipment error*. Even the best scientific instruments exhibit (sometimes frustrating) variations in their functions, and school labs seldom are equipped with the best instruments.

Aside from such observer and equipment errors, problems may arise because of uncooperative or inadequate targets, contamination, inability to control all relevant variables, failure to recognize and control extraneous variables, carelessness, technical uncertainty, poor communications, and chance variations in the circumstances in which we work. All of the frustrations that you as a teacher must deal with in setting up and running hands-on activities are generally found in professional science as well.

In MBST, errors are accepted as an innate part of the process of building a model. You and your students make mistakes, but when mistakes occur that result in a misleading model, you will of course feel obligated to correct them. The best way to do this is by pointing out contradictions either within the model, or between this model and others. No one learns from mistakes that go unexamined. You want your students to learn to look for signs of errors and account for them.

Teachers often impose too much control over work in the lab in order to ensure an error-free positive outcome, but you must be careful that this doesn't lead back to the cookbook labs we mentioned earlier. Following recipes is not a high value skill. Dealing with error is part of the scientific process.

How do we recognize errors? It's not always easy. When scientists make errors they do not recognize, other scientists, through the checks-and-balances of peer review, may catch them. If no one recognizes the errors, the effects will eventually show up when the erroneous model resists further enrichment, or cannot be linked to other accepted models.

Quantitative scientists, meaning the vast majority of natural scientists, also rely upon statistical analysis to reveal the fit of their data to their models. Poor reliability—data on the same variable that do not cluster around a mean—indicates the possibility of errors in the methods used to build the model or to collect the data. Reliable data that do not support our expectations or our hypothetical models may indicate errors in the expectations of models, or problems with our methods.

Engaging Students in Assessing the Validity and Reliability of Their Models

In most traditional classrooms, students conduct their practical work individually or, more commonly, as part of a small work group. Unfortunately, these same individuals or small groups may complete their lab write-ups—in whatever format they use—without knowing how the activity turned out for others in the class. This doesn't reflect real science.

A lab research scientist working alone must collect enough data to convince others that his model is reliable and valid—that it is not just a chance deviation from the norm. His work is reviewed by his peers. He will thus study as many examples of the same phenomenon as is necessary to ensure that he has a predictive, consistent model before he publishes or presents it.

Because students usually don't have time to study multiple examples of the same phenomenon, the best way to mirror the practice of science is to pool our data and use the composite data model as the basis for our individual models.

Pooling data allows us to assess its reliability. For numerical data, the usual rule of thumb is to have at least thirty measurements for any given characteristic, but we can approximate the spirit of statistics if we simply look for a good clustering of data in any set of independent measurements. If we find outliers such as those in Figure 5.9B, we must then decide what they mean.

Outliers skew (distort) the average. The problem is to decide whether the outliers are a real effect or are due to chance or error. The matter could be resolved by collecting more data, but if that's not practical, you must decide on the likelihood that they are errors. If you find there is reason to think they are errors, you may decide to omit them from the calculation of the mean.

While this might seem like "cheating" or "massaging the data," scientists must make similar decisions about their data all the time. They generally reply upon powerful statistics programs to help them in their decisions, though. If you elect to omit data, though, this fact should be mentioned and justified in the final model.

The same basic ideas apply to the analysis of qualitative data. "Averaging" qualitative data is harder but, as in numerical data, you are looking for consistency. The benefit of pooling data is that a group not finding a particular characteristic in their own work is alerted to its presence in the findings of others.

Validity is a subjective judgment of how well one believes the model represents the specific target studied. Have you created a solid representation of the phenomenon?

If in the fourth grade we create a model that compares the rate of fall of objects of different densities (weights), chances are good we will end up with reliable data and a valid model of the target.

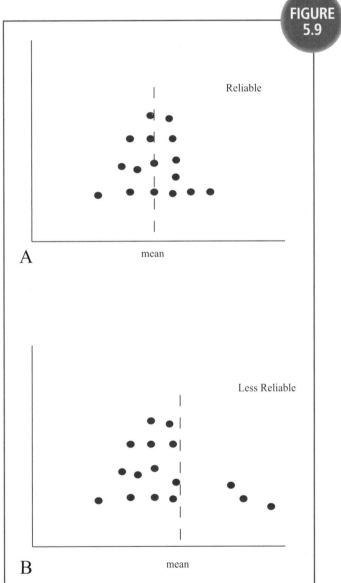

FIGURE 5.9

A   mean

B   mean

Outlying data points can unduly influence the average (mean) value as shown by the shift of the mean from scattergram A to scattergram B above.

While reliability is a quality of the data, validity is a quality of our model as a whole. The question you must ask in assessing the validity of your model is whether the known data (facts) support the model well enough for you to accept it.

Here's the thing: you may collect reliable data that does not support the model. In that case, the model is invalid even if the data are reliable. But the reverse is not true. You cannot construct a valid model using unreliable data.

The question your students must always ask themselves is whether the data are internally consistent (reliable), and then whether the data actually support any conclusions they would like to draw from it.

From Model to Target(s): The Fine Art of Generalizing

The progress of science would be very slow if every scientific model was interpreted as a representation only of its specific target. True, scientists do construct models that describe or explain a specific target, but they also use these accepted models analogically to find explanations for similar targets in related systems.

Generalizing refers to the extension of a model of a specific target to represent other more general targets. For example, the furnace in our basement probably operates in a way that is similar to most other furnaces of its type constructed around the same time. You knowledge of your own furnace can probably transfer generally to other similar furnaces. By the same token, most automobiles with the same power source are similar in basic structure and functions.

There are several types of chlorophyll, but we can presume for general purposes that the process of photosynthesis is similar in all of them.

While we may take it for granted that our students will grasp the generalized applications of the models they build intuitively, our assumption is questionable in light of frequent student complaints about the irrelevancy of science. And if students do not generalize, they lose much of the value of learning science.

Generalization is an important component of science. Scientists generalize as a matter of course. No one much cares about an isolated finding in a simplified system. Run lab rats through a maze and the model of behavior you create is relevant only for those specific lab rats unless you generalize to broader targets: first to all lab rats, then to all rats, then to humans, for example. With each generalization, you must add qualifiers to your inferences because there are more and more differences among the targets.

The models we have our students create in the lab and field must be generalized to have any significant meaning. We intend them to be exemplars, but we select them because they are convenient: the reason we run lab rats through a maze instead of humans. What we end up with is a model that is convenience and economical to build within the confines of the school, but that may not extend, in the mental models of most students, far beyond school walls. But that's also true in professional science, where models may be limited by available resources, complexity, and ethics, among other factors.

So, how do we extend the meanings of our models? By encouraging generalization, of course. Only by generalizing our models can we give meaning to them. As an example, let's look at the elementary teacher who has engaged her students in an MBST lesson on magnetism. Her focus is upon helping her students build a mental model of these concepts, which she will understand through their expressed model. She introduces the concept of magnetism by leading the class to build a model of materials a magnet will attract. She enriches their model by having them add to their model by testing how two bar magnets behave in relation to one another, using good inquiry techniques. Finally, she sprinkles iron filings over a bar magnet and asks the students to speculate on what the resulting lines mean.[5] The students then plan a model to describe how the lines change as two models are moved around one another.

These hands-on activities, taken together, can create a basic mental model of magnetism for most students. Students could generalize their model to all bar magnets, all magnets with other shapes, and to all things magnetic. This suggests other tests that we could make to test the generalization: for example, modeling the lines of force for other magnets with shapes like disks, rings, horseshoes, marbles, and trapezoids, to name a few. Our models get much dicier as we move into electromagnets, electrical appliances (turned on), electrical transmission wires, planets, and suns.[6] You would expect to see essential similarities, but also differences among these "magnets" as you move away from the simplicity of the bar magnets.

Our essential goal for teaching is for our students to understand that the model they build can have applications beyond the classroom, but that generalization entails risks. The further you move beyond your immediate model, the more risk there is of error.

In the example of lab rats in a maze, there is probably little risk in supposing that the lab rats represent other lab rats of the same kind or even rat behavior in general; but supposing that the behaviors of lab rats represent the behaviors of humans is more risky. In some case, though, there is little risk in generalizing to the same kinds of systems in different circumstances. The growth of crystals in a glass of water in a classroom is essential similar to the growth of crystals in geodes, for example.

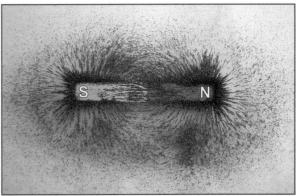

Iron filings tracing the magnetic field of a bar magnet

Photo by Dayna Mason

---

5 The lines are called *lines of force*. They are imaginary. The iron filings line up with patterns in the magnetic field and show us where the field starts and ends. Like most forces, it is otherwise not well understood.

6 All electrical currents produce electromagnetic fields, so any operating electrical device has some magnetic properties—one reason not to leave your credit cards too close to an operating electrical device.

As a teacher, you should lead your students to consider the broader applications of the models they create. Let's take a common cell lab as an example. Biology teachers often have their students construct a model of an Elodea cell. Elodea is a common aquatic plant with large, easily studied photosynthetic cells. In makes a good archetype, but it is considerably different from plant cells adapted for other conditions.

Now, we don't know what those differences might be, but it's reasonable to suppose that that the further removed a cell is from the lifestyle of Elodea, the more differences it will exhibit. So we can construct a model such as Table 5.4 to express our expectations.

Why go the trouble? First, most scientific models have a broader purpose than to describe an immediate target (which is often a sample or a simplified version of the target system). Scientists must identify this large target the model is intended to represent. Second, we want our students to understand that the world is more complex than the simplified archetypes they construct in school science. Third, we would like our students to question claims that appear to overreach the evidence. Doing so is part of being a good consumer of science. It helps if they understand the dangers inherent in generalization. Fourth, we want to model science, and part of the scientific process is to speculate on ways new models might be used to explain a range of systems. And finally, such speculation often results in the realization of a range of potential new research projects. In the case of Elodea, for example, we might be interested in learning how the functions of other plant cells influence the form of those cells, using Elodea as the standard.

Generalization is a key component of human creativity but it is also a source of error when the generalizations are not tested. MBST can help us teach our students to speculate wisely.

Elodea canadensis leaf cells taken at 450x magnification on a wet mount.

The Scientific Research Model

A scientific model is a construction, usually a research paper, which describes or explains a phenomenon according to standards adopted by the scientific community. The model may describe a completely new phenomenon or modify previous models of a phenomenon. Published scientific models generally include the following elements:

- a *clear statement of the problem* and a reason for building the model, usually referencing existing models

- if the model is a test, the *hypothetical models* that are being tested and the predicted outcomes

TABLE
5.4

Confidence With Which a Given Model can Represent a Range of Targets

Targets	Confidence in Representation	Majors Sources of Differentiation (Generally Cumulative)
All similar photosynthetic Elodea cells	Highest	Similar habitats so differences due to regional environmental differences
All photosynthetic cells in water plants		Major differences due to variations in light, salinity, water turbulence, and predators
All photosynthetic cells in plants		Major differences due to availability of water and threat of predation
All plant cells		Major differences due to differences in function and availability of sunlight
All cells	Lowest	Major differences due to motility, food sources and specialization

- a *description of the procedures* used to test or build the model such that they could be replicated by others

- a *data summary or data analyses* revealing important findings and relationships, including an analysis of the reliability of the data, with cautions and explanations for any omissions or modifications

- a written *analysis of the patterns and relationships* found in the data as they relate to the purpose of the model

- an *analysis of the validity, significance, and limitations of the model,* including implications for further enrichment

These are standard elements in most of the working models constructed by professional scientists. Research models are constructed in a variety of formats depending upon the scientific discipline and the standards of the journal or conference where they appear.

In the course of planning MBST activities, teachers must consider each of these elements, specifically:

- What is the problem and reason for building the model?

- What could the model look like?

- What data are most likely to reveal key relationships?

- How should data be treated and interpreted?

- What outcomes are likely, including potential misconceptions?

- How can the model be generalized and limited?

• What does the model reveal about science and the context of science?

• How can the model be enriched?

As we have previously stated, students should be involved in the planning of their models, but as a teacher, you will need to lead this process. The Rates of Fall lesson plan in the text box is an activity suitable for grades 3–5. While the activity as written has good purpose and engages students in hands-on inquiry, its format is not rich: it will not achieve many learning goals and will teach nothing about science as a real activity. Let's look at this activity and remold it by considering each of the questions listed above.

What Is the Problem and Reason for Building the Model?

The introduction in this example is not bad. The teacher raises a problem and involves the students in testing their predictions. However, this is not how scientists operate and is thus misleading. An alternative approach that would be just as effective would be to present them with a tennis ball and a baseball, ask them to handle them, and then ask which will fall faster and hit the ground first if they are dropped at the same time from the same height. Chances are, most will say the baseball (it's heavier).

Then have a student stand on a chair, desk, or ladder (to get enough height) and drop the balls at the same time. Do this repeatedly until the students form an opinion about the balls' relative rates of fall. Now tell your students that they are going to build a model testing their ideas in a more formal way.

What Could the Model Look Like?

The problem is to build a model with a number of different objects to see if the observed phenomenon holds true. Scientists often affirm (or disaffirm) observations that others have made in more informal situations.

In planning the activity, think of the variables you might manipulate, measure, or hold constant as you conduct the research for the model. This will give you the familiarity you need in order to plan the model with your students (and have the materials prepared). Variables you could manipulate include the weight of the objects, their sizes and forms, and the heights of the drops. Variables you could measure include the time it takes the objects to fall or, more simply, observations of their relative moments of impact (i.e., do they hit at the same time or not?). Variables you would want to keep the same would be the moments of release of the two objects and the surface areas of the objects.[7]

You could collect data as a class or in small groups. If you only have one stopwatch, you may want to dispense with timing the drops in favor of collecting and tabulating student perceptions of whether the two dropped objects did or did not

---

[7] Normally important in air only for very light objects or objects with unusually large surface areas.

hit at the same time (yes, no, or not sure). If you want to time the falls, that can be done as a separate activity or you can do the drops as a whole-class activity—however, the downside of this approach is that you will collect less data or have a much longer activity.

What Data Are Most Likely to Reveal Key Relationships?

Regardless of your choice, you must know the limits and parameters of the activity before you can plan with your students. Some of the questions you might use with students to plan the project include:

- What data should we collect and share?

- How should we work and record the data?

- How high should our drops be (the higher the better)

- What objects should we drop and compare?

- How many times should we do a given drop?

Notice that we do not have a preset worksheet. You cannot design a worksheet until you have decided what data to collect. Teaching students how to create worksheets of their own with which to collect data is a skill we want to teach. They cannot learn to do it if you do it for them.

How Should Data Be Treated and Interpreted?

If you are timing the drops, you will need to make the distances of all drops the same in order to find the mean (average) of all times for any two objects. If students are voting on whether the objects hit at the same time or not, their votes will be categorical data that cannot be averaged. Because of uneven releases, you are likely to get some "no" or "not sure" votes along with many "yes" votes. This uncertainty is a cause for experimental error, and you can then discuss how you can allow for it.

What Outcomes Are Likely, Including Potential Misconceptions?

If you are familiar with the principle behind this activity, you are aware that it pertains only to the fall of objects in a vacuum. Objects fall at different rates through

---

TEXT BOX 5.2

Rates of Fall

Grade Range: Any elementary grade

Why Do It? It's common sense to many people that heavier objects fall faster than lighter ones.

Objective: To construct a valid and reliable model of the relative rate of fall of objects of different weights but similar size.

What You Need: Heavy and light books; tennis ball and baseball; heavy and light coins; any other objects that are similar in size but different in weight. Optional: a balance or scale capable of handling weights of these objects, and a stopwatch or stopwatches. Worksheet.

Activity: Have students weigh the objects and record the data. Ask students to rank the objects in terms of how fast they will fall compared to one another, "1" being fastest and ending with the number of objects you have. Have the students work in pairs to test their ideas. Let them time the fall of each object. Make sure objects are dropped from the same height. They may find it worthwhile to drop two similar objects (e.g., tennis ball and baseball) at the same time. Have them record their results and draw conclusions based on their observations.

other media (air, water, or oil) and will be influenced by factors such as the currents and thickness of the media. In the case of air, the influence is small when the distances are short and the objects heavy enough and similar in surface area.

You can avoid the misconception that air has no influence by enriching your students' models with the following demonstration. Take two sheets of typing paper. Make one into a ball and leave the other one flat. Drop both pieces, holding the flat sheet in the middle between two fingers or by the edges, folded without a crease, so it straightens out when you drop it.

Because of air resistance, the flat sheet will fall more slowly. If you then ball it up and drop it again alongside the first piece of paper, they will drop at the same rate. Ask your students to explain the difference and to modify their models accordingly. You can thus address the misconception that the medium doesn't matter.

How Can the Model Be Generalized and Limited?

This model is appropriate to describe all situations in which objects are influenced by Earth's gravity. Would it be true on the moon? On a gas giant like the Sun or Jupiter? We might assume so, but we don't know for sure, since we can't run tests there. Given that gravity seems to work the same way everywhere, we can probably generalize our findings to these other targets, but we would want to keep in mind that there could be exceptions.

What Does the Model Reveal About Science and the Context of Science?

Protoscientists like Aristotle taught that objects of different weight fell at different rates. It never occurred to him, apparently, to test this presumption. This principle in particular appears to violate reason and points up the need to test our "common sense" assumptions.

How Can the Model Be Enriched?

Almost any model can be enriched if time permits. All you have to do is identify variables that could make a difference in the outcome of the experiment. What variables could we modify in this model? Currently, our independent variable, the one we are controlling or manipulating, is the weight of the objects. Our dependent variable, the one we expect to vary by the weight, is the rate of fall (technically, when the objects hit, but that is a product of the rate of fall). Variables we are holding constant include the drop height, time of drop, and the surface areas of the objects. Because extremely light objects would be influenced by the air, we avoid these. There are, of course, many irrelevant variables, such as the color of the objects.

Varying the surface areas dramatically might alter the outcomes. A skydiver with an open parachute will fall much more slowly than one in freefall. A down feather will fall more slowly than a basketball because it is more influenced by air. An object in water or oil will fall more slowly than the same object in air. Drop two items in water and they will fall at about the same rates. Drop a dime in water near boiling and another near freezing and the differences in water density will cause the latter to fall more slowly. Surface area and density of the medium, then are two variables that can be tested and used to enrich this model.

Summary

Berman's quote at the beginning of this chapter implies that simplification is not the same as simplism. Experts in a field often have difficulty explaining a model to novices precisely because they are unable to pare out its essential elements. The expert may find meaning in the details that escapes the novice, who is still trying to get a grasp on the essential elements of the model.

We discussed the values of parsimony, front-loading, themes, and exemplars. These concepts are not new to modern constructivist theories of teaching. To build useful models, we must begin with simple models and enrich them in a parsimonious way by adding to or modifying them. The theme of model building is the essential framework of MBST, but other themes emerge as organizational models for the content we wish to teach. Some of these thematic models overlap and unify the various sciences. They bring focus to our efforts to understand the world as more than just a bunch of unrelated facts and relationships. The goal of science is to construct a grand model that unifies our views of reality across disciplines. We find many of these similar relationships through analogy, but analogies must be validated by testing; never just accepted on their face.

Research reports are the basic building blocks of science. Most scientists prefer to base their models on quantitative rather than qualitative data, because the former can be subjected to statistical analysis yielding definable averages and estimates of reliability. Scientists frequently develop scales to convert qualitative to quantitative data, although the use of scales often requires considerable subjective judgment.

One of the most important skills that a teacher can develop through school science is the ability to collect and analyze data of various kinds. Such data are the heart of scientific models. Students should also become familiar with the concepts of reliability and validity.

We have emphasized in this chapter, and in others, that science is not just experimenting: It is experimenting with a purpose, which is to build a model of a phenomenon with an eye to explaining it. We should include lab and field studies in the curriculum for one major purpose: to teach students how to design and evaluate a model of a phenomenon scientifically. All other purposes for lab work should follow from this one. Students must be involved in planning their models to the degree that they are able to participate meaningfully. They should identify what data they will collect and how they will organize it. They should decide beforehand how they will present that data.

We also must make students comfortable with the fact of errors and teach them how to respond to them. Some errors can be ignored or averaged, while others must be explained away. Some may lead to the rejection of the model. Most models they build will be acceptable. It is appropriate for you, as a teacher, to question the validity of a model and to cite conflicting findings from science, but it is inappropriate to say what should have happened. If there are contradictions, it is better to speculate on why the model might conflict with established science.

Successful school science models generally represent a local and isolated target system that we have set up for our students to study. It is important to draw attention to the limits of the models they build, and to consider how far they may be generalized to describe other systems. Can we claim that all species of plants respond to light in the same way? Even if we can legitimately extend our local model to larger systems (as with gravity), we should do so deliberately, rather than leave it to the students to figure out the larger significance of the models.

Although you must preplan MBST activities, you should try to build in enough flexibility for students to have a say in planning the model they will build. Instructions for activities available on the internet or in published lab and field guides tend to be technical guides, often of the cookbook variety. The good MBST teacher uses these resources only for initial guidance, adapting them in planning and conversations with students that allow those students to learn the practice and nature of science.

For Discussion and Practice

1. Create a lesson plan template that includes spaces to address the concerns identified in the last section of this chapter. Find an inquiry activity and use the template to organize your thinking about how you will present and guide your students in completing their model.

2. You engage your students in building a model of the response of pillbugs (aka woodlice) to bright light. You tell your students you found the animals under a board and wondered whether they went there to avoid light. Having posed a problem, you plan a way to test how pillbugs would respond to light by shining a bright light on them and giving them shade to which to retreat. They do not retreat as expected. How would you handle closure of this model?

3. Your students collect data constructing a model of the boiling points and freezing points of several organic substances. They pool their data and find some measurements are off by up to 10 degrees. How could you handle these data anomalies?

4. Select any activity relevant to your grade and subject and examine the activity closely. Create a mock-up of the model you might expect your students to create from it. Now identify several targets the model could represent (immediate, next most immediate, and so forth). As your targets become more removed from the model, what variables do you find might alter the fit of the model to the targets, making it less useful for representing the target? (Use the pillbug model in item #2 for practice; then try a different activity.)

5. Create a statement in your own words that relates the concepts of parsimony, frontloading, themes, and exemplars. Do not refer back to the chapter. Share your model with the rest of the class or workshop using active listening. Identify ways your model differs from models your peers present.

References

Bertel, S. 2005. Show me how you act on a diagram and I'll tell you what you think (or: spatial structures as organizing schemes in collaborative human-computer reasoning). Paper presented to the American Association for Artificial Intelligence, Menlo Park, CA.

National Academy of Sciences (NAS). 2005. *America's lab report: Investigations in high school science*. Washington, DC: National Academies Press.

Chapter 6

THE CREATIVE PROCESSES of SCIENCE

*The imagination imitates. It is
the critical spirit that creates.*

—Oscar Wilde

Because of the way the term *creativity* is used in popular media, we more often associate it with fields such as art, filmmaking, and writing, rather than with mathematics, science, and the scholarly pursuits. But by this time, you should see that this attitude is wrong. Science couldn't exist without the creative spirit.

By definition, *creativity* is the act of creating something uniquely new, useful, and meaningful. Creativity is not fantasy, with which it seems confused in some peoples' minds. It is a process of imagination, though. *Imagination* refers to the creation of mental images without direct reference to external stimuli. It is necessary, but not sufficient, for creative thinking.

Recall that we create a streaming mental model of external reality while we are conscious (and to some extent while we are unconscious). That's reality to us. It is not considered imaginative because it directly references our perceptions. But if we drift off into a daydream, then we are imagining. Or if we plan and visualize our next move before we have made it, we are imagining.

Imagined realities do not have to be unique, new, or useful. The high school student who dreams of being a physician is not being usefully unique; rather, she is transposing her image of a physician's lifestyle upon her own. The result is pleasurable to her and may be motivating, but it is not particularly creative.

On the other hand, if in her dreams she creates a plan for attaining the goal leading to her becoming a physician, then she is creating a unique product (because everyone's situation is unique) that has use and meaning. This daydream is the product of her *creative imagination*.

But what does Wilde mean that only the critical spirit creates? To most of us, *criticism* means finding fault with something. But it also refers to making skillful judgments about truth or merit. In science, this translates into having the skills needed to identify important problems and address them. An individual lacking critical capacity will not see any need to create anything new because he will not see any problems. Problem *finding* and problem *solving* both require the critical capacity to which Wilde refers.

Science could not progress without finding new *solvable* problems.[1] Nor can science teachers meaningfully engage students in learning to construct scientific models if they cannot identify problems to solve. This is not as scary as it sounds. Problems are all over the place. All you have to do is see them.

In this chapter, we will examine a way to create new problems for your students to solve in the lab and field. We will also look at ways to stimulate our students to think creatively about the meaning and structure of scientific models in general. Creative thinking about our models forces us to know and understand them, and it can be fun. In addition, we want our students to be able to separate the elements of imagination, creativity, and fantasy in their models of how we learn and understand the world.

The Art of Creating Problems

In Douglas Adam's 1979 *Hitchhikers Guide to the Galaxy*, a supercomputer named Deep Thought, after thinking for seven and a half million years, delivers the ultimate answer to Life, the Universe, and Everything. To the dismay of the builders' descendents, the answer is 42. When they express profound outrage at this enigmatic answer, Deep Thought suggests that perhaps their problem is they don't know the ultimate question.

Science teachers (and science professors) spend a lot of time helping students learn the "right" models—the models scientists have already accepted. Too often, they neglect to teach their students how to ask questions. Framing the right question is a creative activity. The question does not exist until you compose it in your mind. And in science, you will recall, our goal is not just to ask the right question, but to create a convincing model that answers the question.

---

[1]　Humankind has always recognized its problems. Finding solvable problems has been the real challenge.

The "right" question is one that is small enough and has enough focus to be answerable. If the system you are testing is too complex—has too many variables to control—you may not be able to construct a meaningful model. Professional scientists strive to create simple but meaningful models.[2]

Part of the art of finding good problems is not aiming too high with any one model. What we want is not only to develop our students' capacity to construct meaningful models scientifically but to motivate them to do so—to feel the satisfaction that scientists feel when their models are respected by others in their field. To do that, they must participate in creating problems as well as solving them. Problems give their work purpose.

Scientific creation begins with solvable problems. For educational purposes, virtually any problem with an unknown solution can lead to meaningful experiences in scientific model building, but the best models will enrich those that already exist.

The problem innate in building descriptive models usually involves recognizing a pattern or a set of relationships that gives meaning to the targeted system. For example, in dissections in biology class, we may look for parallels in the systems we examine, examining how form follows function.

In physics, we may create a model describing patterns in the wavelengths of sounds that cancel or amplify one another. The creation of a telling and meaningful model is, of course, the problem in description.

Experimental science involves *testing variables*. You will recall that most experiments have

- an independent or manipulated variable that you deliberately change, or that has different values you want to compare;

- a responding or dependent variable that you measure to see if differences in its value correlate with changes in the manipulated variable; and

- extraneous variables that you have to hold constant or account for in case they are influencing the outcomes.

Experimental models can often be enriched simply by (a) selecting among the extraneous variables to find a new variable to test or by (b) varying the identity of the manipulated variable. For example, an activity commonly found in elementary science classrooms involves making a lemon battery. In this activity, a steel nail and a copper penny are inserted into separate slits in a lemon. A voltmeter attached properly to the metals will show a current flowing, and if you hook up five or six lemon batteries in series, you can light an LED bulb. It is a very dramatic and instructive descriptive model (no explanation is being tested). How can you create models to enrich it? You can systematically manipulate the variables that are held constant in this model, of course. Students could test

---

[2]   The models may not seem simple to the nonprofessional, but that is usually because of the complexities of unfamiliar concepts and of jargon. We should say simple to the knowledgeable professional. Such models are often referred to as elegant because they explain so much so clearly.

- other combinations of metals;

- other fruits and even vegetables such as a potato;

- current flow in a warm lemon to that of a cold lemon; or

- lemon juice instead of a lemon.

Each set of experiments adds value to the model and enriches it. As a teacher, you can lead students gently to suggest these tests themselves by asking the right questions. Even from this simple example, it's obvious that planning and constructing a scientific model is a creative act.

Science projects that simply teach students how to do something may be interesting, but they do not add value to their understanding of science or their abilities to do science. Examine any set of activities on a science experiments web page or in a science activities sourcebook and you will find most of them are demonstrations students imitate or carry out as if following a recipe.

Students completing a recipe lab are not exercising their creative and critical model building skills. Often these activities end abruptly once the demonstration is complete; and as a science teacher and science teacher supervisor, I have seen many teachers fail to conclude their labs in meaningful and satisfying ways.

We talk a lot in science education about teaching students to think critically, but we seldom follow though and engage them in critical thinking. Consider again the definition of critical thinking we gave earlier. As students plan and engage in constructing good models, they must think critically and creatively.

Analogy, Simile, and Metaphor in Science

You may recall from chapter 1 that an analogy can be expressed as a proposition with a formal generic structure: "A is to B as C is to D." A *simile* is the expression of an analogy as a likeness without identifying what that likeness is: "A is *like* C." A *metaphor* goes further and expresses an identity: "A *is* C." You can see this relationship best if we deconstruct a metaphor:

- Atoms are the building blocks of the universe (metaphor)

- Atoms are like the building blocks of the universe (simile)

- Atoms are to the universe as building blocks are to a Lego house (analogy)

Analogies provide us with creative ways of interpreting the world, and since our mental models are constructed only from the systems that we know, it is fair to say that analogy is one of our major sources of explanation. Analogies are prevalent in every language. Unfortunately, analogies are as apt to lead us astray as they are to suggest legitimate explanations. For example, scientists of the 18th century believed that electricity was a substance akin to water flowing through a pipe—a reasonable analogy that worked well for a while. Electricity is caused by the flow of electrons, but nothing spills out if you cut a wire carrying electricity (unless the live wire is in contact with a conductor, of course). Clearly, the flow is different from water in a pipe. Our current metaphor is more like a bucket brigade[3]: buckets of water passed from person to person that stops when the line is broken.

While we have mostly discussed MBST in a lab or field setting, MBST is most effective if the ideas we have suggested pervade lecture, discussion, and non-lab activities as well. The search for analogies (or similes and metaphors) or, conversely, formulating explanations for existing metaphors, can be fun and productive if that effort includes a critical analysis of the analogy. It can be counterproductive if the analogy is treated as a literal similarity[4] when it is not.

The search for analogies can be as simple as asking students to complete the stem, "X is like…." Discussing similes that students might suggest in response to the stem, "A leaf is like…"can engage your students in critically analyzing the truth of the simile. This adds to their mental model of leaves, as in the exchange below:

S1: *A leaf is like the plant's stomach.*

T: *You mean it grinds and digests food?*

S1: (pause) *It makes food.*

T: *Does your stomach make food?*

S1: *No.*

T: *Does any part of our body make food?*

Ss: *Uh-uh.*

T: *There is something a leaf resembles. Have you ever seen a solar panel?*

S3: *We have one on our house. It makes electricity.*

T: *And you use the electricity for what?*

S3: *To do things with. To make our lights work.*

T: *To do work. And that's what the leaf does. It makes electricity and uses it to make sugars where the energy is stored. The plant can use that energy to do work. Can our solar panels do that?*

S1: *I don't think so.*

Alternatively, you might discuss some terms from an analogical perspective: terms such as magnetic poles, ladder of life, mitochondrial powerhouse, electric current, carbon footprint, drainage basin, carrier wave, weather front, electron cloud, electromagnetic waves, and electron particles are all metaphorical. (For a more complete list of analogies you can go to *www.scienceanalogies.info*.)

---

[3] Electrons pass from atom to atom rather that flow freely like water in a stream.
[4] A literal similarity denotes parts of different systems that are the same in terms of both parts and relationships; in an analogy, only the relationships are the same.

Students often find the concept of an electromagnetic wave confusing, precisely because the analogy suggests something like you might find on a lake. This is, of course, where the analogy came from. Physicists could have called in an energy oscillation and perhaps prevented some confusion. Analogies can be confusing if they are taken too literally and so need to be explored. The question most students have is, "waves of what?"

T: *That's an interesting question. I wish I could answer it. No one really knows what energy is.*

S1: *So how can we make a model of it?*

T: *We can model its behavior and its effects on us and other things without knowing what it is. Scientists and natural philosophers made models of the behavior of sound and light before they could explain them. We know what sound is now, but not so much light. Remember, our models are simplified and useful, but not complete.*

S2: *So why did they call it a light wave?*

T: *More properly an electromagnetic wave. Because it affects matter as though it's oscillating.*

S1: *Going up and down like a water wave you mean?*

T: *Up and down or back and forth or bigger and smaller: a regular increase and decrease in energy intensity that's analogically like a water wave. So they called in a wave. But it could have been called an oscillation. Energy is the ability to do work. Whatever EM energy is, it can do work, but it oscillates at different rates or frequencies and we perceive the energy differently, depending upon how fast it oscillates.*

This kind of deliberate dissection of an analogical model serves two functions: it enriches your own model by forcing you to examine the nature of the target with more intensity than you might have otherwise; and it helps students to understand the nature of knowledge as a model, not as the physical reality itself.

By the same token, you can also make your students aware of the analogical nature of many of the physical or mathematical models you use to teach science. It's important for students to be reminded that the ball and stick models or the electron dot diagrams they use in chemistry are far from completely descriptive of their targets. Asking them to explain what the models represent *and* what they omit goes a long way toward developing their critical understanding of learning as model building. Here's how a classroom diagram might go:

T: (holds up a ball and stick model of a water molecule) *What does this model represent? What's its target?*

S1: *A water molecule.*

T: *At one level, yes. It represents something important about a water molecule. Each part of the model corresponds to a part of the target. Can anyone tell me what those correspondences are?*

S2: *The big ball represents the oxygen atom and the two smaller ones are hydrogen atoms. The sticks are bonds.*

T: *So the balls and sticks of the model represent parts of the target. How about relationships in the target?*

S3: *The hydrogen atoms are bonded to the oxygen—not to each other.*

S4: *The hydrogen atoms are at an angle so it looks like Mickey Mouse.*

T: *A Mickey Mouse model. (pause) Is this what a real water molecule looks like?*

S1: *I doubt it. It's just a model.*

T: *So let's imagine for a moment what the real molecule would look like. We've studied atoms, so you know the scientific model for an atom. What might a water molecule look like in real life?*

S2: *Well, for one thing the oxygen atom would be much larger compared to the hydrogen. The hydrogen atoms are just single protons.*

S5: *The atoms would be a lot farther apart.*

T: *How much farther?*

S5: *Like from the Sun to Pluto or something.*

S3: *And the bonds would be an electron cloud.*

T: *Could you see the bonds?*

S2: *Probably not.*

T: *So if you could look at a real water molecule, what would you see?*

S1: *Probably not much of anything. Like if you looked at the solar system from outside, you wouldn't see much except the Sun unless you knew where to look. Even it would look tiny. It's mostly space.*

By being forced to use their imaginations and to go beyond the concrete model, these students have shown they understand analogy. They also have a much better idea of what the target might actually look like.

Creativity and Conceptual Blending

Analogical thinking often results in *conceptual blending*: the overlap of conceptual mental models to create hybrids combining certain characteristics of each of the blended mental models. This transposition is a source of creative fantasy. If we take a tree and combine with it certain human characteristics, we end up with the Ents (the giant walking trees) in Tolkien's *Lord of the Rings*.

Most of our ideal models, and all of our fantasy models, evolve through the transposition of reality models by us or by others. How, for example, could we even conceive of moving things around only with our minds? Actually, there might be a number of explanations: we summon or control spirits; we change the nature of reality; or we focus invisible energy to do it. Each of these explanations transposes models. The invisible energy explanation for example, is analogous to pushing an object with your breath. Transpose that model onto the situation (moving a penny with your mind) except you are going to "blow" with mental energy (other metaphorical models are also possible). Summoning spirits is just like summoning invisible helpers to do your work. Changing reality is like changing a channel on the TV. The metaphor you accept depends upon your experiences.

As a teacher, you need to be aware of conceptual blending because it is the source of many misconceptions—that is, undesirable creativity on the parts of your students. Because their model may be less stable than yours are, they are more prone to blending. Of course, not all misconceptions are due to blending. Some erroneous models are just errors of logic (low pressure raises liquids when you suck on a straw) or perception (the surface of the Sun is smooth). Table 6.1 identifies several "blending" misconceptions in several fields and the everyday experiential models that lead to the misconception. What happens here is that an existing model based on experience becomes blended with the scientific model.

Blending between seemingly related concepts can also occur because the models have not been developed well enough, or labels are poorly attached to the models. Models commonly confused include force and energy, heat and temperature, adhesion and cohesion, centripetal and centrifugal, magnetic and electromagnetic, virus and bacterium, asteroid and comet, and climate and weather, to name just a few.

We said before that the unfortunate image of science as lacking creativity comes from the confusion of creativity with fantasy. All productive scientists have to be imaginative and creative, but they must temper their speculations with respect for facts. Fantasy is disciplined only by plausibility, for the artists or writers who create fantasy only have to worry about consistency within the story. This is one reason why medieval explanations of the world lasted for so long: they might have been inconsistent with the physical facts, but they were consistent within the world of ideals. If you discounted worldly facts as illusory and temporary (as opposed to the eternal truths of religion), then their worldview made sense.

A scientific model must be plausible within the constraints of existing theory, and that theory must be factual. The model has to obey the rules of nature. An explanation or event that violates accepted laws of physics, for example, such as a superhero flying with no logical means of propulsion would require extraordinary evidence to be acceptable.

TABLE
6.1

Some Common Misconceptions in Science Due to Conceptual Blending

Misconception	Blending Model
The Earth is sitting on something.	Everyday objects must have something under them to support them; therefore the Earth must be sitting on something.
Evolution is goal-directed.	All artists and builders have a goal and purpose for their creations; therefore evolution is directed by a creator.
Dinosaurs and humans lived at the same time.	Movies, books, and cartoons depict humans and dinosaurs together.
When water boils, the bubbles coming up are air.	When you blow bubbles in water through a straw, they make the water boil.
Clouds are made of smoke.	Clouds are just like the smoke that rises from fires, smokestacks, and other sources of flame.
Something has to be constantly pushing an object for it to move.	When you push a car or wagon or box: you have to keep pushing or it will stop; the same is true for all such motion.
Batteries are filled with electricity.	A batter is like a tank or reservoir is filled with water, except they are filled with electricity.
Gases don't have mass.	You can't see gases or hold them in your hand so they must not have any mass: you can see and hold mass.

Infusing Creativity Into School Science

We have already discussed a few ways that we can infuse creativity into school science. Our main concern is to teach our students how to construct plausible, factual mental and expressed models, not just in the lab, but also through readings, discussion, case studies, and other sources of information. Critical analysis is important for creative work: We can only see new and different relationships if we can deconstruct the models that exist.

Engaging in Creative Inquiry

The raw material for creativity is inherent in acts of inquiry, whether in the lab or elsewhere. To *inquire* means simply to learn by asking. The ultimate creative act is to ask original questions and find original ways to answer them. The emphasis on *creative inquiry* (as opposed to informative inquiry in which someone else—an authority—provides the inquirer with the answer) characterizes science.

Creative MBST requires you, as a teacher, to set up opportunities for creative inquiry. This means confronting your students with solvable problems, at one time called *invitations to inquiry*. Guided creative inquiry is used extensively in elementary and middle level science, but it is less common in secondary and university classrooms. The position taken in this book is that all teachers should use MBST to infuse creative learning into their curriculum.

The focus of our creation is on descriptive or explanatory models. Since we have already described this process in previous chapters, we need not go into it again here.

Engaging in Visualization

Visualization is a creative technique for enriching a model. The reason scientific models may seem dead and uninteresting to our students is that the students do not really engage in visualizing what that model represents. Visualization is the technique the teacher used in the dialogue about the water molecule. Visualization is actually an important tool in science: the ability to see in the mind's eye beyond the immediate model. Students who only see numbers and letters in a formula or equation will not fully understand their meanings.

Einstein, Kelvin, and many other scientists spent time visualizing their mental models. Saying that electrons jump from atom to atom as they pass through a conductor is not the same as visualizing this happening in our mental model. This kind of visualization also helps to strengthen and affirm the model; visual images are more enduring for the most part than propositional ones. Visualization is easy, too. Just take the time to ask your students to imagine.

The concept of transpiration in biology is important for the understanding of plants. During transpiration, plants lose water from their leaves through pores called stomata. As water molecules evaporate, an imbalance is created (a "thirst") with the water in the stem and roots. Since water molecules bond to one another ("hold hands"), a chain of water molecules extends from the leaf through cellular "pipes" (xylem) to the roots, where there is more water. Water moves upward in response to the imbalance, to replace the lost water.

By deliberately visualizing this whole process, we impress it upon our imagination as a whole story. You can think of visualization in the context of a thought experiment. What would happen if we broke this chain of water molecules? What if we stopped transpiration from occurring?

Look at a potted plant in your room, or sit and look at a tree on a warm day. Try to imagine the water vaporizing from the underside of the leaves. See the vapor in your mind's eye. See it leave the internal surface of the leaf, leaving room for more water to flow upward through the many thin woody pipes that are only cell thick. See the pipes extending from the leaf through the stem and trunk to the roots where more water flows in from the damp soil.

THE CREATIVE PROCESSES of SCIENCE

Creating Stories

Elementary teachers often create stories to make visualized elements more real—e.g., the story of the little water molecule that gets free through a leaf, only to fall again as rain. Students can write stories too. Scientific explanations are nonfiction stories, told to accord with and explain the facts. You can make science more interesting if you frame science in this way.

Darwin created the story of how the Galapagos finches attained the diversity they exhibit. Each model we build in science is a story. The elements of the plot are added as the model is enriched. Sometimes there are twists and turns in the story, as when the existing form of a model no longer works (meets a conflict) and must be changed. In that sense, all of science is storytelling. The storytelling framework does not conflict with MBST, since stories are by definition mental/expressed word models.

Stories usually have a beginning, middle, and end. Case studies are a type of story we will not deal with in this book, except to say that *cases* are exemplary models developed for dissection and study. *Historical stories* should also be given more emphasis in the science curriculum than we usually give them, but care must be taken to separate these models from the myths that often burden them.

Not all scientific models make good stories, but students could have fun writing and sharing their stories of how sodium married chlorine, for example. The benefit to the students in this case is that they are forced to know and understand the model of ionic bonding.[5]

Brainstorming

Creating explanations is a brainstorming technique that engages students in finding potential explanations for phenomena. Obviously, scientists must create and test explanatory models for unknown phenomena. For example, when Jocelyn Bell Burnell and Antony Hewish observed the first pulsar on November 28, 1967, they called their model LGM-1, LGM standing for "Little Green Men" because the regularity of the emissions seemed to come from an extraterrestrial (ET) civilization.

Pulsars were later found to originate from rapidly spinning neutron stars. While most scientists probably didn't take the ET hypothesis seriously, the implications of such a find were discussed (Burnell 1977) and ET buffs were hopeful for a time that the model would withstand scrutiny. Ultimately, of course, it did not survive in the face of other options.

The search for answers to scientific questions usually begins with the creation of speculative hypothetical models. Over time, these models are tested and the simplest, most consistent, and most explanatory models are usually retained. Scientists may literally brainstorm hypotheses when confronted with a puzzle for which they have no good answer, such as the signal from the first pulsar.

---

[5]   If this strikes you as too sexist, you can reverse the gender roles or just use neutral terms such as "it."

In the classroom, *brainstorming* requires that you collect ideas (hypothetical models) to solve a problem without any intervening critical analysis. The questions you might ask to engage your students in brainstorming are many and varied, but might include:

- What creates mountains?

- Why don't crayfish get huge?

- Why doesn't a fly ball sail on forever?

- Why do elements combine with only certain other elements to form compounds?

- Why are some leaves flat while others are not?

- What makes astronauts float in space?

Science teachers often ask such questions rhetorically, but then go on to provide the answers without engaging students in the search for plausible hypothetical models. From the MBST perspective, this is a mistake.

Table 6.2 presents five plausible hypothetical models for the formation of mountains that your students might suggest.[6] Rather than go directly to the "right answer," you could identify the known facts that would lead them eliminate all but the current scientific model, which is hypothesis five. This would more accurately reflect the nature of scientific creation—of scientific model building.

Interestingly, Benjamin Franklin suggested the gist of continental drift theory in 1782 when he observed:

> The crust of the Earth must be a shell floating on a fluid interior.... Thus, the surface of the globe would be capable of being broken and distorted by the violent movements of the fluids on which it rested. (History of Continental Drift 2010)

TABLE 6.2

	Hypothetical Models for Formation of Mountains
1.	God created mountains at the (relatively recent) Creation
2.	Mountains are created by volcanoes or by volcanic upwelling under the ground
3.	The Earth is contracting and mountains are the wrinkles on its surface
4.	Mountains are lighter material floating on heavier material beneath; eroding material forces new material upward
5.	Mountains are formed when lighter crustal plates move around and collide

[6] Plausible in your students' common worldview models, but not to scientists, since the creationist view (#1) violates a fundamental tenet of science requiring natural explanations.

Franklin's observation shows how many models we take for granted as modern actually were suggested many years ago. What has changed is our ability to collect evidence to support speculative models others created in the past.

What Could Happen If ...

Speculative questions are intended to raise interest and develop flexibility in thinking, thus lessening your student's tendency to think of science as a body of petrified knowledge.

The physical world is always in flux, but change usually happens so slowly that people are unaware of it. For example, the constellations we take for granted are changing as we move through space. The solar system is not as stable as many people suppose either. This kind of thinking can have disastrous consequences, as in the case of global climate change or the insidious effects of pollution. Fictional models such as Clark's 1968 novel, *2001, A Space Odyssey*; Crichton's 2002 novel, *Prey*; or the 1979 film, *The China Syndrome* all depend upon plausible speculation.

The purpose of speculative questioning is to engage students in considering models that may seem outside of the realm of what is normal for them. How would you respond to the following questions?

- What could happen if the magnetic poles reversed without warning?

- What could happen if the Gulf Stream stopped flowing?

- What could happen if we found and brought back a life form from Jupiter's moons?

- What would happen if we discovered how to prevent death caused by aging?

- What would happen if sustainable fusion gave us unlimited energy?

Some of these questions are technological, of course, but some also fall within the realm descriptive/explanatory science. Care must be taken not to dwell on the negative or to pose a situation that might be hopeless, unless you want parents contacting you about why their child can't sleep at night. Situations should be proportional to the capacities and interests of the students.

Summary

Creativity comes about through insights: "eureka!" moments that occur when one thinks flexibly on a problem and arrives at a potential solution. The nature of creative insight is not well understood, but it seems to occur when we transpose concepts, deliberately or subconsciously, to create a novel recombination of nodes and relationships in our analogical mental models. Darwin and Wallace, for example, transposed elements of Malthus's economic model onto the biological model of evolution to create the concept of natural selection.

In contrast to popular perceptions, science is by definition a highly creative pursuit. Scientists only advance in their profession by constructing sound original models that support or challenge established models. To create original models, one must first imagine them. Not all imagination is creative, but it is hard to conceptualize any advances in science that do not depend upon creative imagination.

Scientists combine their creative imagination with critical analysis, which leads them to find problems, see patterns, and make skillful judgments. The best way to engage students in the creativity of science is engage them in inquiry beginning with the creation of a problem and ending with a completed expressed model.

The simplest way to find a researchable problem is to ask questions about completed models. This is what scientists usually do when they begin a new research program. They become highly familiar with existing models in an area of interest and build upon them. With practice, you can become proficient at enriching established inquiry models.

For Discussion and Practice

1. Examine Oscar Wilde's statement at the beginning of this chapter. Discuss it as a class or workshop, or in small groups. What is he saying? Do you agree or disagree? Why?

2. What is the difference between being a scientist and being a scientific thinker? What are the characteristics of a scientific thinker?

3. What is meant by the *myth of the rigorous adherence to logic* referred to in the chapter?

4. Choose two of the ways of stimulating creativity identified above and create a protocol for using them for a lesson in your classroom. Share your creations with others as time permits. Explain how you would proceed and what you would expect to achieve by each lesson.

References

Burnell, S. J. B. 1977. Little green men, white dwarves or pulsars? *Annals of the New York Academy of Science* 302: 685–689.

History of continental drift—Before Wegener. 2010. *The geology of Portsdown Hill. http://www.bbm.me.uk/portsdown/PH_061_History_a.htm*

Chapter 7

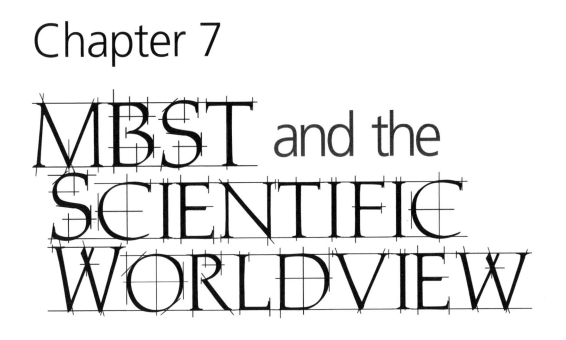

MBST and the SCIENTIFIC WORLDVIEW

I should venture to assert that the most pervasive fallacy of philosophic thinking goes back to neglect of context.

—John Dewey

To understand anything fully, you must know and understand its operational context. Context is everything. A joke told among young men at a fraternity party might not go over among a group of elderly churchgoers, and jokes that make sense in one language and culture may go completely flat in another one. Likewise, to attain scientific literacy, you must understand science within the framework of its psychological, social, historical, and cultural contexts.

Most of our contextual mental models are nonscientific and a few may be unscientific (contradict scientific models). We develop all of our mental models through experience and adapt them through trial and error to create a practical worldview—one that enables us to interface with our environment reasonably effectively from our own perspective. But we seldom test our personal models systematically in a scientific manner. For example, we may accept some explanations because others in our social network have adopted them, because people we regard as experts profess them, or just because they fit in with our own brand of logic. We don't always need evidence to support our views. Intuition may be fine for some purposes.

We all operate with our own, unique worldview model; each of us is the central player in that model. Although no model is identical from person to person, we normally share core elements of some of our models with others in our society and culture. Cultures evolve when members of a large population share core models that distinguish them from other large groups. Subcultures are distinctive groups within a given culture. Members of the predominant scientific subculture across societies tend to share certain common values of science that have evolved over the last three and a half centuries (Table 7.1).

TABLE 7.1

Elements of the Traditional Scientific Culture

Traditional Scientific Culture:
Assumes all perceptible phenomena to have a logical discoverable cause
Values predictive models that lead to fruitful, stable explanations
Values tested factual explanatory models over speculation and analogy alone
Values open exchange of information and peer review
Makes its decisions and rules based on facts and validated models
Spurns models based only on dogma, intuition, and traditions
Recognizes the value of intuition and speculation as a source of ideas to be tested
Neither accepts nor denies the possible existence of hidden worlds, forces, or deity
Recognizes that science may not be able to explain all things

In this chapter, we will outline the elements of a model of science in several operational social and personal contexts. Once we have developed a model for these contexts, we will look at ways you might ways might contextualize science in your classroom. Let's begin by examining the scientific worldview more closely.

Defining a Model of a Scientific Worldview

The whole of science is nothing more than a refinement of everyday thinking.

—Albert Einstein

No one exactly agrees upon what should be included in a model of a scientific worldview. Certainly, the individual with a predominantly scientific worldview differs in some respects from those who hold a strictly technological worldview (an engineer, for example) or any one of the many other worldviews that might be found in any society, a sample of which are presented for illustration in Table 7.2.

The scientific worldview is not exclusive to the professional scientific community, nor do all of the members of the scientific community share the scientific worldview in the same way. For our discussion, we will assume that the individual having a scientific worldview should exhibit all or at least most of the traits identified in Table 7.3.

TABLE 7.2

Comparison of Several Predominant Worldview Models

Worldview	Greatest Interest in	Focuses on
Scientific	Explanation	Knowing how systems work in the perceptible natural world; explaining things on a factual basis
Technological	Invention	Engineering solutions to practical problems; Less concern with why something happens than with how to make use of it.
Political	Power and governance	Sees the world in terms of power models and how to use models to control others and attain desired ends
Social	Human relationships	Focuses on models of interactions between people and the dynamics of how people relate to one another to achieve their personal goals
Intuitive	Spirituality, feeling, aesthetics	Quest to find meaning through feeling and analogies as opposed to empirical testing and systematic analysis
Hedonistic	Pleasure, sensual gratification	Quest to find pleasure through sensual gratification
Economic	Generation of wealth	Making money through trade and commerce, increasing social and personal wealth through the making, buying, and selling of goods

I doubt that any of these traits would generate raised eyebrows among most science philosophers and educators. One that might is skepticism, because it is sometimes associated with doubts toward religion. However, we will define *skepticism* in the broader sense of a reasonable questioning attitude toward claims and models that others may accept and even take for granted. Skepticism may also be confused with *cynicism*, which is similar expect that it implies a questioning of human motives. As science educators, we may want our students to be skeptical, but most of us are not interested in creating habitual cynics.

Unless they are hermits, no one living in modern society can avoid the numerous claims of scientific endorsement found in ads and arguments in the media. Many of these claims are unfounded, or ought to be questioned properly, at a minimum. The person with the scientific worldview is in a better position to determine the truth of

TABLE 7.3

Elements of a Scientific Worldview Model

Traits	Explanation
Analytical	Desires to know how things work and why things are
Open-minded	Knows that all models are subject to change with good cause
Factual	Adheres to fact as the primary source of knowledge
Impartial	Does not bias data to favor certain conclusions
Logical	Connects the dots without leaping ahead
Empirical	Assumes any perceived phenomenon has a discernible cause
Realistic	Does not confuse speculation and fantasy with tested reality
Skeptical	Questions any claims made without evidence
Undogmatic	Rejects dogma and doctrines of any kind
Practical	Creates sane, sound, brass-tacks models

such claims, especially if he or she recognizes the communication as a simplified entity designed for a purpose. The first thing most scientifically literate people will look for is purpose; then they will examine the underlying facts and assumptions. Many claims are not factual. They are based on people's willingness to trust authority. Without facts, claims can only be accepted provisionally, at best. And if the stakes are high, one would be well advised to be cautious in accepting claims without solid evidence.

Characteristics of a Scientific Society

Most societies and cultures are not highly scientific in their collective worldview, but some cultures are more compatible with science than others are. Few societies today actively oppose science as a whole, but they may limit the work that scientists can carry out in certain areas of inquiry. These limits may not be matter of law but social pressures, job security, and funding restrictions are often used to maintain social, cultural, or political restrictions on certain kinds of models.

An ideal scientific society would by definition value scientific models and support the free dissemination and discussion of scientific models by its citizenry. The scientifically oriented society would probably exhibit all or most of the characteristics shown in Table 7.4.

In reality, of course, every culture includes many models that are unscientific and nonscientific, and most of their people do not possess and act on a scientific worldview. In fact, stories that embody cultural ideals, myths, legends, and traditions shape many of our ideas. Science is, after all, a means to an end. Some would argue that it is not an end in itself and, taken alone, cannot result in a truly fulfilling life. We all need some fantasy, they say.

TABLE
7.4

Traits of Scientific Societies

Trait	Justification
Strong secular educational systems	Scientific thinkers value learning unrestricted by religious dogma.
Free speech and information exchange	The uncensored exchange of ideas and information is essential to productive scientific thought.
Democratic political systems	Scientific thinkers generally thrive in a non-repressive political atmosphere that respects the individual's right to think as he or she pleases.
Freedom from repressive dogma	Dogma is the enemy of scientific thinking because it places some areas of inquiry off-limits.
Freedom of inquiry	Scientific thinkers value their right to inquire into any issue they choose without being unduly limited by others.
Social tolerance and stability	Instability tends to limit free thinking and models that induce uncertainty while tolerance is necessary in order to consider potentially unpopular ideas.
Rules and laws informed by science	Rules and laws based on tested predictive explanatory models are generally less repressive than those based on blind tradition, guesswork, religious customs, and analogy.
A strong science tradition	A culture of scientific thinkers is likely to support and promote science as an ongoing institution to the extent that their resources allow.

We will not address this issue here, since philosophers have failed to resolve it despite centuries of debate. It is questionable whether anyone's worldview is strictly scientific; however, some worldviews are more scientific than others are. To understand the scientific worldview, we must compare and contrast it with other opposing worldviews. Chief among these is the mythical worldview model.

Science and the Mythical Model

People say that what we're all seeking is a meaning for life. I don't think that's what we're really seeking. I think that what we're seeking is an experience of being alive, so that our life experiences on the purely physical plane will have resonances within our own innermost being and reality, so that we actually feel the rapture of being alive.

—Joseph Campbell

No individual or society, no matter how scientific, lives without myths. That myths are an essential component of normal human thought cannot be denied. They are prevalent and useful—as long as they are recognized as myths.

Traditional *myths* are fictional stories with largely symbolic actors—human or nonhuman—used to explain phenomena (e.g., creation myths) or impart lessons (e.g., the myth of Prometheus). *Legends* on the other hand are unverifiable or fictitious stories with the quality of folk tales and sagas. In some cases, myths and legends may be popularly accepted as true.

Legends often have mythical qualities, and the terms *legend* and *myth* are frequently used interchangeably. The elements of myth are shown in Table 7.5. Myths are *not* simple misconceptions and errors in thinking—a use of the term that is common but wrong. The symbolic quality distinguishes myth from other stories.

We recognize some myths and legends as fictional on their face. They may be too fantastic to be anything other than metaphorical tales; no knowledgeable person regards *Lord of the Rings* or *Star Wars* as more than works of fiction. The symbolism in these stories imparts to them a legendary, larger-than-life quality lacking in much generic popular fiction.

We may value myth for many reasons. Myth, as Campbell and Moyer (1991) say, appeals to our innermost sense of being. In the mythical worldview, the need for internal consistency more powerfully determines what is real and important than the need for consistency with facts. Strong mythologizers may feel it is more important to maintain the integrity of the internal story than to adapt to the reality of the senses. They may dismiss the logical or scientific objections of others to their model as unreliable or biased. Certainly, this describes the typical medieval mindset, and it describes the mindsets of many people today.

Traditional myths are stories from the past, but we used the term today to refer to any story—any explanation—that is proposed and retained because of its innate symbolic appeal, despite a lack of evidence supporting it.[1] The *Five Stages of Grief* model promoted by Elizabeth Kübler-Ross (1969) is one example. According to the model, dying patients pass through five stages of grief: denial, anger, bargaining, depression, and acceptance. This model is often referenced by counselors and ministers, despite any lack of scientific support for it (Friedman and James 2008). While these responses may show up, they may not occur in any particular order or even at all. Anger and bargaining, for example, depends upon concepts of a deity that not everyone shares. Some dying patients never accept their own deaths.

But the model has a mythical status because it symbolizes the human struggle with death, ending with the desired stage of acceptance. It is neat and tidy. In fact, it is too neat and tidy, if you think about it. Given the enormous variation in human worldviews, can it be that the act of slowly dying can be summed up so neatly?

---

[1] Simple erroneous models that are not myths lack symbolic appeal.

TABLE
7.5

Traits of Myths and Legends

Trait	Explanation	Example
Symbolism	Many myths have a symbolic or metaphorical meaning that is greater than might be expected from an historical event. That is, there is a lesson or meaning embedded in the story. Most stories do not attain legendary or mythical status.	There is no evidence that Galileo mumbled "and yet it does move" after he recanted heliocentrism. The statement is symbolic of defiance but appears to be part of the myth of Galileo.
Few or no facts	This distinguishes myth and legend from history. Myths may begin with a real event at their core, but the elements of myth that are added are nonfactual and unsupported by reliable sources.	We have no facts to verify that God created one sex to be superior over the other. Rank is situational and an assumption of deliberate creation is not factual.
Magical or supernatural	Magical and supernatural phenomena are typical of certain classes of myth and legend.	Miracles are often the stuff of myth and legend, but none have been shown to have a basis in fact, nor has any supernatural event ever been demonstrated to be factual.
Exaggerated	Myths and legends often make claims that seem to supersede normal life boundaries. They have the feel of composed drama rather than life as we normally experience it.	The heroes of most myths rarely engage in common life tasks or make normal mistakes. They are often imbued with exceptional qualities that set them apart.
Unscientific	Myths often contain elements that are unscientific.	
Illogical and irrational	Some myths contain elements that are illogical from the perspective of what we know about a situation.	The notion of an all-powerful timeless deity can be shown to be logically impossible on several levels.
Promotes a cause	Myths may be manufactured or come about to promote a cause, often feeding on prejudice, cultural norms, or personal gain.	The Frankenstein myth is often used to symbolize anti-science sentiment, although Dr. Frankenstein might better be viewed as a biotechnologist.

Here's another example of a common model with mythical status. As an educator, you may have heard that we learn 10% of what we read, 20% of what we hear, 30% of what we see, 50% of what we see and hear, 70% of what we discuss, 80% of what we experience, and 95% of what we teach others. Pretty neat, huh? The problem is that there is no published literature that backs up this story (Genovese 2004). It is a common myth among educators that has the superficial appearance of science (numerical and systematic) without factual substance. It is accepted because it is widely cited but no substantive research backs it up. Writers and speakers (and teachers) cite other writers, speakers, and teachers without ever questioning the source, which is how such myths evolve. If you look carefully at it, you realize it is too neat to be real: the numbers are too neatly multiples of five or ten, and a 95% retention rate is literally unbelievable. Like the Kübler-Ross model, it has simplified symbolic value; in this case, as a "scientific" argument for active learning.

We value myths because they provide us with order and explanation, even when we recognize them as fictitious. They are motivating, especially when they project ideals to which we can aspire. These ideal models are often—in fact, are usually—simple and easily assimilated into our mental models. They appeal to our parsimonious natures. They are usually consistent with what we want to believe, providing us with models of right and wrong, good and bad. Myths provide us with heroes and villains. They justify our actions and beliefs. Shared myths may give us cooperative unity and strength. But, of course, they can also be created and propagated to manipulate and exploit us.

The fashion diamond industry, for example, actively constructs two myths: (1) the myth that enduring romantic love can be bought with a rock and (2) the myth that diamonds are rare. And one of the most successful and enduring advertizing campaigns of the 20th century—that of Coca Cola—was built upon the mythical association of this soft drink with patriotism and American pride. I am not arguing that there is anything innately wrong or evil in marketing through myth; only that consumers may suffer harm if they buy into some myths without really understanding them as such.

Urban myths may seem at first glance to be odd or unusual stories. They become myths because they aptly and simply symbolize certain fears: of technology, government, science, religion, the insane, or just about anything we can fear. They are more than mere false stories.

While myths can lead us to make bad choices, they also are the sources of many desirable cultural beliefs and norms. In the United States, for example, the Declaration of Independence and the Constitution have acquired mythic qualities. In the minds of many people around the world, they are symbols of revolution and human freedom. The simplistic notion that they "enshrine" principles of freedom speaks to this mythical quality. It prevents us from tampering with these documents lightly. But we can respect these documents on a more reasoned, factual level as well. Myth is not necessary.

Idealized differences between mythological and scientific worldviews are outlined in Table 7.6. Most people, of course, have blended worldviews, not pure types. It's impossible to check (or even to know) the facts about everything, so error is inevitable, even in the worldview of the most conscientiously scientific skeptic. We are human, after all, and the tendency toward mythologizing is supported by the very nature of our processes of mental modeling.

Science and Supernatural Modeling

Beliefs in magic, superstition, the paranormal, and the supernatural originate in our mental models of a world with which we can communicate and that influences our own (Table 7.7). The supernatural worldview is common across societies and cultures. For the time being, we will distinguish it from the religious or mystical worldview that we will discuss next, although there are obvious parallels.

Lindeman and Aarnio (2010) propose that superstition, magical thinking, and paranormal beliefs originate in the overlapping of core mental models we develop in childhood that allow us to distinguish animate from inanimate objects, and to relate intention (purpose) to animate objects. Like all mental models, these structures sometimes overlap. When they do, the confusion of attributes lays the groundwork for paranormal beliefs and magical thinking in adults.

TABLE 7.6

Standards for Comparing Mythical Worldview With Scientific Worldview

Standard	Mythical Worldview	Scientific Worldview
Factuality	Few or no empirical facts are needed to support a given explanatory model	Requires supporting empirical facts that can be verified to support an explanatory model
Plausibility	Explanation may require events that are implausible or illogical by everyday standards	Explanation must be plausible and logical in an everyday context
Analogy	Looks to analogy as explanation without supporting evidence	Insists upon empirical evidence to support an analogical explanation
Parsimony	Simplified, and may include magical, supernatural, or covert forces for explanation	Does not reference magical or supernatural forces or allow for covert forces
Testability	Accepts claims that are untested or unaffirmed by disinterested parties or that cannot be tested	Prefers claims that are tested or affirmed by disinterested parties
Truth values	Finds truth in metaphor, symbolism, and spirituality	Finds truth in the physical world
Modes of inquiry	Analogizing, fantasizing, intuiting	Describing, testing, theorizing

TABLE 7.7

Concepts Related to the Supernatural

Concept	Meaning
Magic	The transfer of occult forces between objects, between people, or between people and objects, or the art of controlling supernatural agencies or the forces of nature through incantation, spells, charms, conjuration, prayer, or other means
Occult	Means hidden; usually refers to covert magical or supernatural powers or agencies
Supernatural	Pertains to anything that is above or beyond what is natural, i.e., perceptible. Supernatural worlds and forces are those held to exist beyond the senses
Superstition	The belief in and fear of supernatural forces

In a series of tests, the authors compared the beliefs of superstitious individuals with beliefs of skeptical individuals. Their results, shown in Table 7.8, supported their hypothesis. Avowed believers in feng shui, astrology, and extrasensory perception exhibited higher attribute confusion, intuitive thinking levels, and slightly lower analytical thinking.

While attribute confusion helps explain how magical thinking[2] happens, it does not identify a cause for magical thinking—why some people think magically and others do not (or do so to a lesser degree). Table 7.9 identifies traits of a generic magical worldview model that encompasses superstitions and magical/paranormal beliefs.

Magical beliefs are hard to refute conclusively because they are based upon the assumption that hidden, often willful, powers exist; but since they are hidden and willful, they cannot be scientifically studied. Most of the agents in this occult world are analogous to agents in the physical world. The physical world interacts with the hidden world, but in an unpredictable and willful way.[3] Manifestations of occult agents tend to be ambiguous and the data supporting magical models are unreliable. No scientifically sustainable model of a supernatural cause for paranormal phenomenon has been reported in professional journals by reputable, professional scientists.

Therefore, we cannot prove or disprove supernatural phenomena except through argument, which is no proof at all in science. Neither can we prove that our car has a mind of its own when it refuses to start up on a cold winter morning. We refuse that explanation because it does not fit into our model of natural cause and effect. Our model of how a car operates does not include any provision for a mind.

Magical explanations depending upon conceptual blending can evolve into a surprisingly stable mental structure, usually defended by the assumption that

---

[2] For simplicity, we will use *magical thinking* to stand for supernatural, paranormal, and superstitious models as well as magical ones.

[3] There are exceptions, as when ghosts, demons, or mysterious lights appear from the supernatural domain.

TABLE
7.8

Beliefs of Superstitious Individuals Compared to Skeptics

Compared to Skeptics, Superstitious Individuals Were Significantly More Apt to
Assign physical or biological attributes to mental phenomena, e.g., giving thoughts physical force on independent life
Hold that the mind can physically affect objects, e.g., change events by thinking about them
Hold that an evil thought can contaminate an entity, e.g., bad thoughts alone bring about calamity
Assign mental attributes to inanimate things, e.g., a car literally has a mind of its own or chooses to break down
See random events as having purpose when they were affected by the event, e.g., a chance encounter with an old friend was meant to happen

TABLE
7.9

Model of Magical vs Scientific Worldviews

Component	Magical Worldview	Scientific Worldview
Explanations	Some events have occult explanations	All knowable events have physical explanations
World dimensions	Assumes that occult powers and agents exist	Does not assume that occult powers and agents exist
Control	World can be controlled by certain combinations of words, phrases, or numbers, or by objects with occult powers such as charms	Control in the world comes from manipulation of natural systems; words and numbers are just symbols
Teleology	Inanimate objects have goals and purposes	Goals and purpose are traits only of living things
Mind force	The mind can move objects and control events	The mind can only control bodily functions
Contamination	Evil can be transferred between objects	Evil is not a transferable essence or substance
Life essence	Life has a substance or essence apart from physical matter (soul)	Life is a result of interactions within physical matter
Cause	Physical events can happen without a physical cause	All physical events have a physical cause
Destiny and fate	Our destinies are shaped by forces external to ourselves and our environment	Our destinies are shaped only by our actions and the environment
Dominate mode of model building	Intuitive, symbolic, analogical	Sensory, concrete, factual, literal

facts are elusive and malleable—that perceptions of physical reality are untrustworthy, and that intuition is a better guide to truth than the sense. This idea is far from new, of course. It has existed in various cultures for millennia.

To maintain the integrity of their models in the face of scientific skepticism, believers may denigrate scientific explanations or processes, arguing that scientists argue from their own biases and reject evidence that is not in accord with their own views. However, its own literature shows that paranormal investigations tend to be shoddy and unconvincing at best and fraudulent at worst.

Our current scientific models do not explain everything, however, and many mysteries remain to be explained. Could it be that an occult world exists? Yes. Could there be occult forces beyond our own? Again, the answer is yes. But whereas the individual with the magical worldview believes that such a world exists and that it can be known through intuition, the individual with the scientific worldview is more skeptical, arguing that intuition is not a firm standard of evidence.

Magical thinking reveals itself in the traditional explanations offered by some for therapeutic touch and acupuncture, among other "alternative medicines." Leaving aside for the moment the question of whether these practices are effective, the explanations offered by traditional acupuncturists for the technique's effectiveness have no scientific basis. Similarly, at least some practitioners of therapeutic touch talk about energy fields that have not been shown to exist. The explanations are speculative at best, but practitioners tend to treat them as explanations, reflecting an analogical, intuitive worldview, rather than a scientific one.

Bear in mind that techniques may be effective without being explained. Ancient physicians and wise women had potions that worked but explained their effectiveness by referring to spirits. Early medieval metalworkers explained their work as interactions of the spirits of metal, air, and fire. Their magical explanations did not stop them from creating many useful products, but they did not gain the benefits of science either.

The confusion of science with magic and the supernatural is common in popular culture. Science fiction, for example, is shelved together with fantasy in most bookstores. While this eliminates the need for booksellers to delineate fantasy from speculative science—a tricky business sometimes—it also erroneously associates science with the supernatural.

Science and Religious Models

We spoke in an earlier chapter about the historical conflict between science and religion. This conflict emerged over the centuries and has more to do with specific unscientific religious beliefs than with theology as a whole.

Religion refers to sets of specific beliefs about the cause, nature, and purpose of the universe and divinity. Most religions have authoritative writings, rituals, customs, myths, and worshippers. *Theology*, in contrast, refers to the broader study and analysis of matters related to divinity. Most organized religions contain strong elements of myth and supernaturalism. Mostly grounded in tradition, their stories have strong symbolic and metaphorical overtones with little grounding in empirical fact.[4] Religion models tend to be built upon analogical reasoning. From a scientific perspective, their lack of a factual foundation renders them nonscientific or unscientific and speculative.

Religious models endure because they cannot be conclusively disproven. The fact that they also cannot be proven is of less importance to adherents than the social cohesion and sense of personal worth and meaning that they provide. In many cultures, religion is the basis for law and order—more so than the secular authorities are.

While scientists can question certain religious phenomena, they cannot disprove the existence of God or an afterlife. The exclusion of God from scientific explanations stems from a desire to (a) remove religious explanations from scientific discourse, and (b) free scientists from religiously imposed constraints.

Some writers, both critics and supporters of science, have taken this to mean that science champions atheism. However, atheism is a personal choice: not a scientific one. Those with a scientific worldview do not reject anything conclusively unless they have facts to disprove it; nor do they accept it. The scientific community in general takes no position on such things as God and the afterlife because it cannot legitimately do so.

Current scientific models of biological and cosmic evolution suggest that evolution occurs through natural processes. The scientific model does not include notions of conscious design or purpose in nature, other than that which is internal to living things within the system.

But the scientific thinker also does not reject the *possibility* of divine interventions conclusively. The scientific worldview does not allow for a closed mind. To paraphrase the fictional detective Sherlock Holmes, whatever is not impossible is possible. Our mental models are always simplified, incomplete, and subject to change. Although we accept scientific models because they work, that does not mean they are complete.

Table 7.10 outlines the significant differences between a generic religious model and a model of science. As a science teacher, you should be thoroughly familiar with the differences in these worldviews, keeping in mind that not accepting a belief is not the same as judging it false or wrong. It could be there is not enough evidence to accept it. Scientific thinkers are comfortable with the notion that they do not know everything and that perhaps that there are things they cannot know.

---

[4] Virtually all religions are based on analogy alone and teach through unsubstantiated stories, thus they are strongly mythical by definition.

Science, the Media, and Informal Experts

An Associated Press article in my local newspaper recently trumpeted the headline: ALCOHOL MORE DANGEROUS THAN HEROIN, COCAINE, AND METH. Now that got me to sit up and take notice!

It turns out that the report, based on a study published in the prestigious British medical journal *Lancet*, found that heroin, crack cocaine, and methamphetamine are more lethal to individuals, but the total damage caused by alcohol individually, socially, and economically is greater than for the lesser used hard drugs.

Assuming that this model is valid, it is hardly the earthshaking story that the headline promises. In fact, it seems to affirm what most of us know intuitively: that the widespread use of a less dangerous drug will be more damaging in total than

TABLE 7.10

Comparison of Religious and Scientific Models

Characteristic	Religious Model	Scientific Model
Purpose	Symbolic and metaphorical explanation	Seeks working models of natural systems
Other worlds	Assumes worlds exist beyond those that are immediately perceptible	Deals strictly with the natural world and makes no assumptions about hidden worlds
God or divinity	Assumes a god or gods, or some form of divinity or superior intelligences exists	Does not presume or deny gods or superior intelligences
Mode of knowing	Depends upon divine revelation or inspiration of prophets and messengers	Depends upon describing and testing causal relationships in the perceptible world
Interventions	May presume direct regular or periodic interventions by a divinity in human affairs	Presumes world functions mechanistically without direction or interference by a divinity
Life after death	Most presume that humans have an incorporeal soul that continues beyond death	Does not assume or deny the existence of a soul or life after death
Nature of truth	Most assume an immutable truth in the words of prophets, holy documents, rituals, etc.	Truth is an alignment of fact and theory that may be altered with new evidence
Ethics and morals	Relies upon revelation from a divinity for ethical and moral standards	Relies upon human reason, logic, and welfare as the only certain source of ethics and morality
Social nature	Primary mode of interaction is community faith, ritual, shared stories, and prayer	Primary mode of interaction is peer review and formal presentations in various forums

the more restricted use of considerably more dangerous drugs. The headline is misleading but is likely to draw in casual readers. Therein lies an apt metaphor for the media worldview. By *media*, I am referring to mass media: magazines, newspapers, radio, television, films, books, internet websites, Twitter, and similar systems for the general dissemination of information to many people at once.

Media outlets depend on an audience of listeners or readers, who in turn have a choice among many outlets or the option to remain uninformed. It seems likely that even the town criers of old walked a fine line between providing truth and providing entertainment. One can only imagine pioneer travelers bringing news to lonely settlements in the old west, embellishing their stories to both foster their own views and to thrill their audiences, the better to curry favor and rewards from the locals.

Media reporters seek what Freedman (2010) refers to as *resonance*: harmony between the needs and expectations of their audience and the stories they report. The most salable stories are those that resonate most with the audience, reinforcing their prejudices, raising their expectations, or, at least, satisfying their needs. In the rush to get resonant stories into print (or into whatever medium they are using), writers, reporters, talk show hosts, and tweeters may overlook details, such as any confirmation of the validity of their stories.

In the eyes of the public, media celebrities and science writers are what Freedman refers to as informal experts. *Informal experts* are individuals who acquire perceived expert status in a field without having the prolonged or intense experience, practice, and education normally associated with true expertise. Informal experts popularize models created by true experts, but often lack the cautionary perspectives the latter have. While the true expert might say, "X is true but you have to also consider Y and Z," the informal expert may simply promote the idea that X is true.

Many writers, teachers, reporters, physicians, talk show hosts, therapists, politicians, commentators, and others may be perceived by the public to have a level of expertise they do not in fact possess. As a science teacher, for example, you may well develop a true expertise in science education, but you are not likely to be a true expert in the specific issues of science you present in your class. But your students may perceive that you have such expertise, and may accept your position on an issue as being the "true" one. This leads to the temptation to preach rather than teach.

Informal experts are necessary for an education society to flourish. Problems arise, though, when, under pressure to perform and lacking the in-depth perspectives of true experts, they present "highly debatable judgment as fact through their choice of which data to highlight" (Freedman 2010)

Informal experts are often well educated. They may even be scientists who are simply commenting on matters outside of their fields of expertise. Physicians and engineers, nurses and therapists may gain a voice with the public by virtue of their degrees, status, and presence rather than their expert knowledge. Michael Crichton, a physician turned science fiction writer, wore the mantle of scientific expertise when he criticized global climate change advocates in his novel, *State of Fear* (2004), despite the fact that he has no special expertise in the subject.

True experts are not immune to exaggeration and self-promotion, but informal experts often make their reputations and livings by taking positions on debatable issues. The positions they take are often those that resonate with their audience, rather than those that are most consistent with the facts. By projecting confidence, they are able to connect with their audience and "do quite well advancing exotic, logic-defying, hard-evidence-free ideas" (Freedman 2010).

Most informal experts probably think of themselves as honest practitioners of their trade. They believe in the positions they take. But the true expert is aware of the limits of his expertise and qualifies the positions he takes accordingly. He does not allow himself just to guess an answer, nor does he make up evidence that sounds right. If he is uncertain about a model, he acknowledges it. He is careful not to distort a model by substituting intuition for fact.

Table 7.11 presents a number of the reasons why informal experts in science might be wrong. As a science teacher—an informal expert in science—you should be aware of these traps and avoid them. You should teach your students about these traps as well. They are common in many fields, not just in science.

The Scientific Worldview and Professional Science

If proponents of the scientific worldview had to worry only about the confusion sown by informal experts, it would be bad enough. Unfortunately, they also have to be concerned with the failings of modern science itself. According to an analysis by Ioannidis (2005), "the facts suggest that for many, if not the majority, of fields, the majority of published studies are likely to be wrong."

This is a rather disconcerting claim for teachers who promote (and believe) one of the cardinal myths of science: that dedication to truth, careful procedures, and peer review make published scientific models almost error-free. Alas, such does not appear to be the case.

The very success of science may be leading to this internal failure—if, indeed, Ioannidis's model is correct. Science has become a huge, impersonal, competitive, and often money-driven enterprise. While its reputation depends upon its internal integrity, scientists face many pressures that may lead them to cheat. Ostensibly, science polices itself through peer review and by replicating studies. Such practices are supposed to catch mistakes and expose instances of misconduct. In fact, replication studies are rare and at least some studies of the peer review process reveal evidence of neglect, carelessness, preferential treatment, and conflicts among reviewers.

TABLE
7.11

Reasons Why Informal Science Experts (ISE) May Be Wrong

Reason	Explanation	Impact
They do not understand science.	Though they may be well-educated, ISE's are not always well prepared in science and may attribute qualities to science and scientists that are not valid. They may not fully understand the processes of science and scientific validation and confuse science with technology.	Dissemination of misunderstandings about science perpetuates misunderstanding and unrealistic expectations, as well as blame for technological misadventures.
They do not fully understand the relevant scientific model(s) or they misinterpret it.	An ISE may understand one model better than another or may not be aware of alternative models that have arisen to challenge the original model.	ISE's are perceived experts; errors they disseminate are often taken up by others depending upon them for guidance. They may then be led to make wrongful decisions based on these errors.
They have an unconscious bias.	The ISE may select among competing models those that support her own biases and "common sense." A strongly religious health professional cites research showing prayer can help cure cancer without mentioning research that refutes the claim.	An ISE with power may have enough impact to cause unsupported models to be accepted by the public. When the model later proves invalid, the public blames science.
They lack skepticism.	The ISE may not know how to ask questions about newly reported models or may be inclined to accept such models unskeptically.	An unskeptical ISE is more likely to jump on a claim without evidence that it has been vetted properly and spread it to others. Later invalidation or lack of follow-up has a negative effect on public perceptions of science.
They corrupt a model for personal advantage.	An ISE may alter the substance or meaning of a real model in order to tell the audience what they want to hear or create a situation from which they can profit. They commit fraud.	The deliberate corruption of models is subtle and often undetected. Such models do not produce the promised results for those who act on them. Science gets blamed.
They pander to an audience.	In this case, the ISE does not directly alter or distort the model but, instead, only presents those models the audience is likely to embrace.	Employees may pander to bosses by only presenting models the boss wants to hear. The omission of negative models, of course, may have undesirable consequences later.
They create their own scientific models.	The ISE must be perceived as an expert and so constructs a model intuitively that seems consistent with those he/she knows.	This fraud is widespread in the media, as well as in professions where degreed and credentialed ISE's pass off their own intuitive models as being based on science.
Ethics and Morals	Relies upon revelation from a divinity for ethical and moral standards.	Relies upon human reason, logic, and welfare as the only certain source of ethics and morality.
Social Nature	Primary mode of interaction is community faith, ritual, shared stories, and prayer.	Primary mode of interaction is peer review and formal presentations in various forums.

TABLE
7.12

Reasons for Scientific Failure

Reason	Explanation
Use of the wrong surrogate (proxy) markers	A surrogate marker is a variable that is *assumed* to stand in place of a variable (such as longevity) that cannot be measured conveniently or easily. The problem stems from assuming a link between surrogate marker and the targeted variable. If you choose the wrong surrogate marker, any conclusions you draw about the targeted variable are void. This is a common problem in medical studies.
Sloppy measurements	Sloppy measurements may happen if assistants are careless or are poorly trained, an instrument is not validated or is not properly calibrated, or for some related reason.
Skewed samples	Choosing samples that are convenient rather than using those that might produce the best outcomes often results in skewing. Using all male subjects for a medical study generalized to both sexes is one example.
Animal substitutions	Many studies on lab animals have produced results that did not generalize to humans, who were the actual targets. This is a broad example of the danger of using surrogates in studies. In some cases, researchers have created models using genetically altered mice, which had no comparability to the actual targets of the study.
Proving a flawed model	Scientists may take failures to validate a model to be a failure in technique rather than a failure of the model. They may then adjust their methods until they validate their model, but the model would not exist except in the artificial situation they create.
Tossing out inconvenient data	All scientists tweak their data, especially to compensate for variations in conditions and variables they cannot control. However, the rules of science do not allow for throwing out data just because it is inconvenient. Some scientists do anyway.
Changing the goal to suit the data	Research begins with a model in mind; sometimes with hypotheses. In some cases the data do not support the hypotheses. One solution is to find hypotheses after the fact that the data do support. This is considered cheating, but there is good evidence that it is done more than just a little.
Confounding variables	Some studies may not control variables that could account for the effect noted in the subsequent model. Instead, the variables are ignored without comment. This problem is especially common in observational and epidemiological studies, but they can also occur in studies comparing controlled groups.
Juggling results	The data are either misanalyzed or are analyzed inappropriately to get a publishable result.
No agreement on variables	This problem occurs with meta-analyses, in which a scientist constructs a model by pooling a number of outcome models focused on the same phenomenon. Since the outcome models may not target the same indicator variables, or control for the same variables, the choices made by the scientist conducting the meta-analysis may unduly influence the final model.
Sabotage or camouflage	This occurs with human subjects when they provide misleading information deliberately, either because they resent the research or because they are suspicious of the uses to which the data will be put.

Scientific modeling may fail for many reasons, some of which are provided in Table 7.12. A disheartening number of failures stem from the inadequate performances of scientists themselves, while some arise from blatant misconduct. For example, a 2003 study undertaken by the American Medical Association found a strong correlation between industry sponsorship of research and positive findings. An anonymous survey of 3,200 funded National Institute of Health researchers published in *Nature* in 2005 found one-third of the researchers admitting to at least one act of misconduct.

Some of the reasons scientists may be tempted to engage in misconduct are shown in Table 7.13. All have to do with the professionalization of science and the modern pressures to publish or perish, or, in many cases, to obtain external funding or perish.

This is not to say that all scientists engage in bad science or that we cannot accept the majority of scientific models. Some scientific misconduct is minor. Bad models get weeded out eventually because they do not work, but in the meantime, they can create confusion and result in wasted time and resources amounting to millions of dollars per year (Harmon 2010).

The good consumer of science is cautious about accepting major new findings until they have undergone adequate confirmation. For the most part, important claims get more scrutiny from other scientists than minor ones. They will be vetted sooner.

The good consumer should also scrutinize the credentials of the authors or publishers of doubtful claims they find on the internet and elsewhere. One must question the claims of scientists who work for, or are funded by, the businesses or industries favored by the claims. They should also beware of "fringe science," which masquerades as science, but with an agenda, and often without the quality control of mainstream science. Fringe journals and organizations may look and sound like mainstream journals and organizations until you look more closely at their products.

SourceWatch (*www.sourcewatch.org*) identifies the Oregon Institute for Science and Medicine as one such organization. It provides a detailed analysis that is an object lesson in how science can get derailed by the personal beliefs of talented scientists. The nonprofit SourceWatch is an excellent resource for teachers seeking information on issues.

So the question becomes, how does the average consumer of science distinguish between good and bad studies? How can they know which studies to accept and which to reject? The fact is, they can't; at least, not with any certainty. One must maintain a skeptical attitude. Table 7.14 presents a number of "yellow flag" conditions. While any one of these conditions would not indicate a model was invalid, they should raise the consumer's caution about strongly advocating the model until the issue is resolved.

**TABLE
7.13**

Major Reasons for Misconduct in Science

Reason	Explanation
Employer expectations	Scientists who work for private employers or the government are often under pressure to produce findings that favor the product or policies of the employer. Padding scientific reports is amazingly common. Another alternative is to suppress any findings that are not favorable.
Competition and status	Some scientists have a strong competitive streak. Teams of scientists may compete with one another for priority (which can bring tangible benefits). The temptation to adjust data more than is warranted to beat a competitor may be overwhelming for some people.
Grants and funding	Grants are expected in many universities and highly valued in others. Universities take a share of most grants from external agencies and many big universities earn the majority of their revenue this way. In some institutions, professors must earn all or part of their salaries through grants. The enormous pressure to get grants provides a big incentive for a little chicanery.
Publication bias	Publication space in reputable journals is limited, and rejection rates for manuscripts may be as high as 90%. Journals prefer publishing positive findings. Negative findings or replication studies are seldom published. Therefore scientists are under pressure to produce significant positive results.
Tenure and promotion	Most universities expect research publications for tenure and promotion, but not all PhD's are able researchers, especially just starting out. The assistant professor facing ejection after six years (or, where there is no tenure, ongoing pressure for publication) may resort to creative enhancements of his or her findings to push publications through.

Developing the Scientific Worldview in the Classroom

Most advocates of science see nothing wrong with calling nonscientific models into question. But teaching anything more than the facts—teaching students how to think and to be skeptical—becomes a political act. Socrates died for it. Most science teachers are not quite as dedicated as that! Yet we also pay lip service to fostering critical thinking. Critical thinking, by its nature, carries the risk of contradicting the valued models of others and thus of offending their beliefs. This is one reason why some teachers and school districts focus so intently on the technical elements of science (content and technique) and avoid its contexts.

TABLE
7.14

Indicators for Caution in Accepting Scientific Models

Indicator	Explanation	Example or Elaboration
Contradicts established models	Any model that contradicts existing models should be held in abeyance until it can be validated.	Example: Regular drunkenness found to lengthen lifespan
Shows cause for bias	A heavily funded research is suspect; research funded by a special interest such as a business or political association is *really* suspect.	Example: More carbon dioxide in the air will benefit the economy, according to a study by the National Petroleum Institute.
Has a little too much resonance	Models that are consistent with our biases or desires and appeal to us intuitively should be accepted with care. The same resonance may bias a researcher.	Example: Most people have a positive attitude toward the future says study (this may be true, but the comforting nature of the conclusion makes it suspect)
Is a simple survey or observation	Surveys depend heavily upon subject attitudes and the way questions are asked, which can shape outcomes.	How was attitude measured? Who was surveyed? Self-reports are notoriously unreliable.
Is unusual and unexpected	A startling new finding should be taken with a grain of salt. Because it is startling, it may have been reported prematurely.	Example: Cold fusion produces energy in a bottle according to Utah researchers.
Lacks peer review	Some models are reported before they have been peer reviewed or vetted by the scientific community. Be careful of them.	This was true of the discredited cold fusion study referred to above. Later reviews and attempts to replicate the model failed.
Seems unlikely	The model seems incredible on its face.	Example: Cancer can be cured by aspirin says NIH researcher. Probably not true.
Has too small a sample	Large samples do not ensure validity, but small samples are especially likely to yield insufficient or biased data.	The possibility that differences are due only to chance increases with small samples.
Report is vague or has no data	The study does not give any indication of the degree to which an effect is seen or gives only percentages without absolute numbers	Is the effect large or small? Is it statistically significant? Be very wary of percentages alone. One is 50% of two.
Uses a skewed sample	Any sample that does not represent the full target population should be considered suspect.	Only men are studied in a model generalized to all people.
Uses a proxy target	The thing measured is not the actual target of interest. The inferred link may be poor to nonexistent.	MRI measures of brain activity are used to infer specific thinking processes.
Is only an animal study	Many animal studies do not yield results transferable to humans. Regard any model constructed from an animal study with caution.	Research shows up to three-quarters of positive drug trials in animals fail or are unsafe in humans (Kola and Landis 2004).
Is correlation only	Correlation does not prove causation. This is one of the most common errors. Other variables might cause the effect.	Change in the price of onions in Paris is correlated with change in the price of steel in Germany. Both are linked to a common factor but do not cause one another.
Cannot be replicated	A study of one event that cannot be replicated or verified independently	A single report of a mysterious phenomena that cannot by its nature be confirmed or replicated

While it is not a panacea, MBST offers one remedy for this by framing all beliefs as models. Advocates of science are sometimes offended by the notion that science is a product of faith, but it is.

"But science creates models that work," they argue. "We can see they work."

"True," I would reply. "But science is explanation, and we cannot see many of the phenomena we are trying to explain. We cannot be sure our explanations are complete or have universal applications. We cannot prove there is no world beyond this one that gives rise to what we think we know. Science is based in faith on the cause-and-effect nature of the physical world, and while that faith seems to be justified, it cannot be shown to be absolute."

As a teacher, you do not need to denigrate models based on myth, the supernatural, or religion. A better idea is to examine them; to understand them; to engage in critical analysis. What role does the model play? Where did it come from? Why is it important to us? What are the core elements? What assumptions underlie the model?

As we've said many times, models have a purpose. We don't just learn anything and everything. That would be wasteful. It would be maladaptive. The scientific approach is to try to understand and explain *why* a model is maintained, even if it seems implausible.

If your students are going to become scientifically literate—if they are to adopt a scientific worldview—it will be because they decide to do so on their own. No one can make them become scientifically literate. Your role as a science teacher is to lead, not to compel.

The role of the elementary science teacher is to develop the concept of mental and expressed models. Upper elementary students should be able to distinguish scientific, nonscientific, and unscientific models. You should take care to clarify to your students that a nonscientific model is not wrong by that fact, but an unscientific model is always wrong from a scientific perspective.

Stories are models, and most models you build in the elementary grades, whether in science or other subjects, can be conceptualized as stories. Stories in science are descriptive or explanatory models based on facts. Stories in literature are often fictional and some are mythical. Some are horror stories that reference the supernatural. You can teach children how to distinguish between these different kinds of stories by calling attention to the purpose and assumptions underlying them.

Around Halloween, you could engage students in speculations about ghosts and spirits. Where do they come from? How do you explain them? How do you prove them (or disprove them)? Is there factual evidence for them? Is there any evidence other than hearsay? Is the idea of ghosts scientific, nonscientific, or unscientific?[5]

Such discussions allow children to think about what is real and what is not. How do we know what is real? What kinds of things can be verified and what cannot? If Harry Potter flies, what is it that moves his broom? If a magician makes something vanish, where does the mass of the thing go? Vanishing is unscientific unless you can create a model that accounts for what happens to the matter. Is flying on a broom scientific or nonscientific? Engaging students in a search for speculative explanations of even the most unlikely things can be fun, but it can also help students realize the difference between guessing and actually having an explanation that can be tested. List the students' explanations on the board and separate them into groups: untestable and implausible; plausible but untestable; testable but implausible; and testable and plausible. The search for testable explanations is, of course, what science is about. Keep in mind that you are not searching for correct answers.

Students in the upper middle and secondary grades are able to conceive of science as model building, and by high school, they should be able to relate the approach to modern conceptualizations of mental models. Case studies are appropriate with which to address scientific integrity and the history and nature of modern science. At the very least, students should be provided with minds-on experience with models that allow them to become familiar with the ordinary and usual tenets of science (Table 7.15).

These tenets can be useful for deciding whether a model is scientific or not, keeping in mind that the fact that a model is not scientific does not mean it is wrong. In the United States, at least, biology teachers are often faced with the delicate task of differentiating the traditional Biblical creationist model and the scientific model of evolution through natural selection. This places them in the position of having to contradict the firmly held beliefs of some students and parents (or their own beliefs, in some cases), or avoiding the topic altogether. Since the more recent iteration of Creationism is Intelligent Design (ID), we will focus on that model.

Some teachers may find it hard to refute a model they, themselves, believe in. It probably won't hurt to reiterate that science may never resolve issues of ultimate reality. What scientists are saying is that there is no *evidence* for ID, and so ID has been left out of their model for the time being. Since they cannot find evidence of divine intervention, they prefer to leave unanswered questions in their models open rather than conclude that a higher power is involved. The day may come

---

[5] Science cannot disprove ghosts, so the idea is nonscientific. Nor does science incorporate ghosts into its models, since they cannot be verified. Current science cannot explain a mechanism for the existence of ghosts, but that fact per se does not directly contradict any scientific model. Nonscientific models need not be believable. On the other hand, if a material phenomenon contradicts a material principle of physics, it is unscientific: material flight without a material source of propulsion, for example. Ghosts may be nonscientific but people flying on brooms is unscientific. Ogres are nonscientific although we commonly acknowledge them as fantasies because they could exist somewhere and their existence would not contradict any scientific laws.

TABLE
7.15

Commonly Accepted Tenets of Science

All Good sscientific Models
Describe and/or explain events in the physical world
Are grounded on facts and logical inference
Are simplified and incomplete representations of their target, subject to change
Are testable and may be proven wrong
Are validated through their consistency and predictivity
Use only natural forces to explain phenomena
Are peer reviewed and considered plausible by experts in the field
Are as free from deliberate bias as possible

when this changes, but for the moment, ID fails to satisfy the ordinary tenets of science in several ways (Table 7.16).

A model is different from an idea. The term *idea* says nothing about the nature of an idea, while the term *mental model* implies attributes, including the attribute of artificiality. Something *artificial* is not false, rather it is literally "made by the arts" of humans. Once students accept that their mental models are things that they create, rather than things they find or acquire, they may feel more freedom to examine them critically.

Summary

Among the many contextual factors that can interfere with, or negatively modify a student's perception of science and scientific models are the models of myth, religion, and the supernatural. Such models are inconsistent with the models of science, although they may not contradict them directly. The assumptions underlying many of these models may not be falsifiable. So while science does not accept the models, it also cannot disprove them. The usual course of action is to remain silent. Specific details of nonscientific models may be shown to be unscientific. For example, we no longer regard mental illness as possession by demons. Such a belief would be unscientific because it would conflict with the findings of neuroscience. Creation of the Earth in seven days is also an unscientific model, rather than just a nonscientific one.

In addition to these belief-based models, contextual factors such as the influence and goals of the media, the culture of informal experts, and the operations of modern science itself have to be understood by any who consider themselves to be scientifically literate. We cannot understand science if we only study its models

TABLE
7.16

Comparison of Models of Intelligent Design With Natural Evolution

Scientific Models:	Intelligent Design	Natural Evolution
Describe and/or explain events in the physical world	Yes	Yes
Are grounded on facts and logical inference	No. There are no factual data supporting intelligent design, only intuition.	Yes
Are simplified and incomplete representations of their target, subject to change	No, the model is not subject to change.	Yes
Are testable and may be proven wrong	No. Intelligent design cannot be tested or proven wrong.	Yes
Are validated through their consistency and predictivity	No. Intelligent design cannot be validated since there are no data supporting it.	Yes
Use only natural forces to explain phenomena	No. Explanation requires a Creator/Designer of the world.	Yes
Are peer reviewed and considered plausible by experts in the field	Yes	Yes
Are as free from deliberate bias as possible	No. Intelligent design is a model intended to promote traditional religious views.	Yes. The current model evolved to replace ID models of the past by following the facts.

and its experimental procedures, or its overall operation in a theoretical sense. We also must understand how it operates in the real world—the world in which we all live. Too many of the models scientists create today prove to be of little value, or to be outright wrong. Not only does this waste millions of dollars in research funding, but it also sows confusion among scientists and the public.

A big part of the value of the overall science literacy model inherent in MBST lies in teaching students to be skeptical of any scientific models reported in the media or proclaimed by informal experts; or any scientific models that have certain characteristics often associated with overgeneralization, deliberate bias, and research errors.

For Discussion and Practice

1. Nonscientific models can be consistent with, or informed by, science—or not. Create a model of some important element of your worldview, such as your concept of family. Using Table 7.5 as a guide, estimate how much of your view of the family is mythical and how much is informed by science.

2. Identify three socially significant issues relevant to your science subject area (or to science in general, if you are not a specialist). Analyze both sides of the issue. How much are the arguments for or against based on factual data? Myth? Special interests? What factors might cause bias in the scientific models relevant to the issues?

3. Examine a news website, looking for announcements of any new scientific model. Read the report carefully, looking for indications of how the model was constructed. Compare your findings to the indicators in Table 7.14. How much confidence can you place in the report based on your analysis?

References

Campbell, J., and B. Moyer. 1991. *The power of myth*. New York, NY: Anchor Books.

Crichton, M. 2004. *State of fear*. New York, NY: Harper.

Freedman, D. H. 2010. *Wrong. Why experts keep failing us—and how to know when not to trust them*. New York, NY: Little, Brown and Company.

Friedman, R., and J. W. James. 2008. The myth of the states of dying, death, and grief. *Skeptic* 14 (2): 37–41.

Genovese, J. 2004. The ten percent solution. Anatomy of an education myth. *Skeptic* 10 (4) *www.skeptic.com/eskeptic/10-03-24/#feature*.

Harmon, K. 2010. Scientific misconduct estimated to drain millions each year. *Scientific American*. *www.scientificamerican.com/blog/post.cfm?id=scientific-misconduct-estimated-to-2010-08-17*

Ioannidis, J. 2005. Why most published research findings are false. *PLoS Med* 2 (8): e124. *www.plosmedicine.org/article/info:doi/10.1371/journal.pmed.0020124*

Kübler-Ross, E. 1969. *On death and dying*. New York: MacMillan.

Lindeman, M., and K. Aarnio. 2010. The origin of superstition, magical thinking, and paranormal beliefs. *E-Skeptic*. *www.skeptic.com/eskeptic/10-09-08*

APPENDIX 1

Student Readings

The readings in this chapter are primarily appropriate for high school and, perhaps, for some advanced middle level students. While it is important for elementary teachers to use the language of models with discussing science with their students, students at that age will probably not be ready to consider the nature of science from a deeper social, philosophical, and historical perspective. It might also be helpful for teachers learning MBST to read these readings as a way of revisiting the major points I've made in the seven chapters here.

My purpose in including these readings is to provide materials through which students can be directly introduced to MBST principles. The material in these readings mirrors the material in the chapters, but without as much detail. The readings are short and concise. You have the option of adding to them. They are intended to stimulate discussion rather to be learned by rote.

The divide between our subjectively constructed reality and the objective world lies at the heart of true science literacy as it is framed by MBST. Having grown up in a world that relies so heavily upon computers and technology, the concept of mental models will not be strange to them. We will have accomplished something special if we can just get our students to contemplate the likelihood that their knowledge of the world is approximate and incomplete, and that they are wise to be flexible in their thinking.

You should feel free to copy and distribute these readings for educational purposes in your classroom.

What Is a Mental Model?

Obviously the facts are never just coming at you but are incorporated by an imagination that is formed by your previous experience. Memories of the past are not memories of facts but memories of your imaginings of the facts.

—Philip Roth

Physical modeling is a uniquely human ability—at least as far as we know. You know what a model is, of course: everyone does. It is something you create to represent something else. But have you ever considered how a model works?

Models work through analogy. An *analogy* shows the same relationship between two things, A and B, that exists between two other things, C and D. We would express this analogy in this way: A is to B as C is to D. When we create a model, we create (or choose) the A-B system to represent the C-D system, our *target*. We create analogies all the time in. For example, what do you mean when you call somebody—let's call him Bob—a star?

Do you mean that Bob is a hot ball of glowing gas? Of course not. You mean that Bob shines like a star: that he stands out and is noticeable, like a star in a black sky. That's analogy. The complete analogy would read, "Bob is as noticeable on the football field as a star is noticeable at night." That's too awkward to say, so we usually just shorten it to "Bob is a star." This shortened form of the analogy, where Bob is same as a star, is called a *metaphor*. If we said, "Bob is like a star," we would be using a *simile*. Metaphors and similes are just shortened forms of analogy. The star in the night, in this case, is a model for some quality of Bob.

You use many models in your life. If you read music, you are following a model. If you read a book, you are following a model. Models aren't just the three-dimensional physical constructions we use to represent other physical things. A *model* is any system that represents another system in a different medium. A book contains a model. A picture is a model. The simulation games you play are models.

Models are always simplified. They never fully represent their targets, because if they did, they would be the target. The reason we build models is to capture some characteristic of the target in a simplified way, so you can never fully understand a target by studying a model of it. Usually you don't even come close. Reading about a thing is never the same as experiencing it.

Why is this important? Because cognitive scientists—scientists who study how we learn and know things—have learned a lot in recent years about the brain and how it works. They don't know everything of course because all they can do is construct a model of the brain and how it works. But computers have turned out to be an apt metaphor for the brain. By treating the brain as a computer, scientists have come up with many new and interesting insights into thinking and learning. And since science is an approach to learning, we need to look at what they know to understand science and scientists.

We can start with the logical assumption that all you know about the world comes from information you receive through your senses. Your sense organs receive energy from the outside—what we will call the *objective* or physical world—and convert it into electrical impulses that are carried by your nerves to the brain. There the information received is processed and combined to create what we might call a *streaming mental model*. This is the world that we are consciously or subconsciously aware of, our *subjective* reality.

Imagine a camera receiving light, converting the light to electrical signals, and sending the signals to a computer where a program reconverts the signals to the picture that you see on your computer screen. Are you looking at reality? No. You are looking at a reconstructed model of reality. And if you are receiving sound as well, it has to be added in through another processing system. What you have in front of you is simpler, less detailed, than the objective target.

I may enjoy watching a football game on the television in my own home, but I cannot get the full experience of actually being at the game. That's not a bad thing. It's just that models are simplifications of their targets. Everything you learn at school is a simplification of the thing it represents.

The reality of our existence, as far as we can tell, is that we can only know these internal streaming models of our environment. We can never perceive the objective world fully and directly. We live in a world of approximations, ignorant of the true nature of existence. That doesn't mean we are creating objective reality in our heads. Scientists assume that an objective world exists. But we can only know it though the models our brain creates. For example, your brain adds color to the world. Color is not an objective property of matter. The processes of your brain—your *mind*—give meaning to everything you perceive. One way to think of this is that your mind is creating a story as you interact with the world. Your mind, of course, is who you are.

Parts of our streaming mental model are stored, consciously or unconsciously, as memories for later use. We can think of these as *reference models*. If I asked you what you had for supper last night, you could probably tell me, but if I asked you what you had two weeks ago, you probably couldn't say. Our reference mental models are unstable. We often forget details and only remember the "big picture" ideas: the *principles* and *images* that guide our decisions and actions. Over time, our stories may become confused. That's probably because our brain has to recreate the stories when we need them, just as your computer may recreate a file from information stored in different places on the disk. Our brain is not a perfect computer, but its flexibility is what makes it special.

Learning is adaptive. With some exceptions, our minds don't retain reference models unless we think they might be of some use to us. We don't necessarily make that decision consciously. We may remember things we would like to forget because the thing has made a powerful impression on us.

We can never know whether our mental models are true to objective reality. The key question is whether they work: whether they allow us to operate effectively in the world. If they work, we accept them as truth, *truth* meaning that our perceptions of events in the physical world are consistent with our explanations and expectations.

No two mental models are the same, because no one has the exact same response to experiences. As you read this paper, your mind is actively constructing a mental model representing what is in it. You are creating meaning in the context of your reference models. The model that you end up with will be different from anyone else's. All mental models are unique. This has some important implications for understanding science and learning that we will explore later in other readings.

For Discussion

1. Eyewitness testimony has come under fire as being more unreliable that was previously thought. How could you explain this using the model of knowing presented in this paper?

2. If scientific models are always incomplete and simplified, how do you know they really represent objective reality?

3. How can we claim to know objective reality if everyone creates different mental models of it?

Student Reading # 2

More on Mental Models

All our ideas and concepts are only internal pictures.
—Ludwig Boltzmann

No one knows just how our mental models are stored in our brain, but we have reason to think that the pieces of the models are not all stored in one place. Different areas of our brain appear to process emotions, sounds, sights, and so forth. Our stored mental models have to be assembled when we need them. Even our streaming mental models—those we are aware of right now—have to be assembled from information from different sources. Cognitive scientists don't know how this is done, and the mystery of how we are aware remains to be solved.

Except for the blind, we humans depend strongly upon visual imagery—on mental pictures. Other animals may rely more on other senses, such as the sense of smell or touch. We store mental images from all of our senses. That's what allows us to recreate a song in our minds, or vividly recall a taste or touch. These images are *sensory images*, as opposed to the *abstract images* we create in our minds without sensory input through *imagination*. Images are probably the most primitive and basic components of the mental models of all reasoning animals.

Within and among these images are networks of *associations* or relationships. You can maneuver around your school or home because of these stored images and their associations. Navigation in highly familiar settings may not even require conscious thought. Such images and associations allow you to *operate* in many situations, not just get around.

If you think right now of a friend's face or voice, you will get a mental image, but it will probably not come to you clearly. If you imagine what your school cafeteria looks like, you can probably glimpse bits and pieces of it in your streaming mental model. Most imagined models don't appear to us like a picture on a television screen. They lack detail. For the most part, they remain in the background, allowing us to operate on them as needed, but keeping a low profile otherwise.

Our minds operate through a network of images, associations, principles, and rules that govern our actions and desires. Principles and rules are just complex associations. All animals with a brain and nervous system create associations, of course. Most animals appear rely on sensory images and simple associations in their thinking, although they may be capable of some abstract thought.

Humans differ from most other animals in being able to communicate their mental models through words, numbers, and symbols arranged in an order with its own meaning. The meaningful ordering of words, numbers, or symbols is called syntax. A statement with syntactical meaning, like this sentence, is called a proposition. Each word has meaning, and the order of the words has meaning as well. That's syntax.

Mathematical equations also have syntax. If you rearranged their parts, it would change their meaning. The songs of birds and whales have *song syntax*, but each note does not mean anything. Only humans seem capable of propositional model building.

The human mind uses sensory images and propositions to express its mental models, both internally and externally. We don't know how these components are stored, but the computer analogy seems to fit. The picture or sentence that appears on your computer is stored in a file on your computer's disk as bits of information—ones and zeros. It is reassembled when it is needed. Our brains probably work in a similar way, except that the information is stored in the nerve cells of the brain.

Sensory images and their associations are the core elements of our mental models. Our ability to process words and numbers syntactically evolved on top of these core elements. That is one explanation for why images are so powerful: why one picture is worth a thousand words. Mental models that include imagery are more likely to be understood and retained than models presented with words alone—assuming that the models are good, and assuming that the learner pays attention to the models. That's why some people find it useful to sketch out ideas.

Some people may find formulas, equations, and numbers alone very hard to relate to because they are abstractions. Numerical processing in the brain is separate from the processing of words. Some people find it useful to explain formulas and equations to themselves in words to make sure they understand what each part means. The act of converting models from one form to another requires understanding that can aid learning.

To construct any kind of a model, whether in science or in some other area of study, it is often more helpful to show than to tell. Words are merely symbolic representations of real parts of your model. If you take the time to visualize the things you study—especially abstract models—you will learn and recall them better than if you do not.

For Discussion

1. Does this model of learning help you to clarify how you think and learn? Is it any different from your previous model of learning?

2. Do you agree that information about the world only enters our brain through the senses? If not, where else does information come from?

3. Models are inherently incomplete and simplified versions of their target. Using the ideas in this reading, explain why.

4. This reading only talks about the brain. Is the mind different from the mental processes of the brain? How?

Student Reading # 3

Science Is Conceptual Modeling

Words, words, words! They shut one off from the universe. Three quarters of the time one's never in contact with things, only with the beastly words that stand for them.

— Aldous Huxley

Words are key elements in our mental and *expressed* models, the latter being any external representation of a mental model. When you speak or write a sentence, you are building a syntactical model. Each word stands for an object, relationship, or characteristic. The relationships among the words correlate with relationships in the objective world. In a simple sentence like "John loves Jill," every word has a meaning and the order of the words expresses a different relationship than "Jill loves John."

Each word represents a *concept*, which is your total idea of a person, place, or thing. In the sentence above, "love" has certain connotations. I could get some idea of your conceptual model of love by asking you to write down everything you can think of that is associated with it. These associations characterize your idea of love. No two people would write exactly the same things and, in fact, even closely related people might express far different models.

Conceptual models are the bricks and mortar of our mental models. Much of what you are doing in school is delineating and labeling new ones. You delineate conceptual models through their relationships with other models. It's instructive to watch a very young child learn. When he first learns the term "mama," he may apply it to all women. As his mental model matures, he learns that mama only refers to one person, but he may resist the idea that there are other mamas in the world. Gradually he comes to understand that mama can refer to any woman with a child as well as his own specific mother.

This process of model maturation occurs no matter what age we are, and what we are learning. For the sake of communication, we agree upon the labels we give our models. "Dog" represents the same animal for all English-speakers. The label "2" evokes the same thing as "two" for most civilized people.

These labels and symbols represent *phenomena*, things that we perceive in the physical world, as well as things we conceive of and wish to represent. They are not the things they represent, and this is important to remember. We don't truly know something if we can't conceive of the model and target behind the word or symbol. One reason some students have difficulty with equations is that all they see are the symbols.

The ability of the human mind to create models from words is truly amazing. No computer can emulate it. Computers work much faster, but almost all programs use algorithms to function: that is, they follow a set of fixed rules. The brain is able to learn and retain information in the form of images and it can operate upon those images. It can create a model out of words to describe those images. Furthermore, it creates its own rules as it goes along. No computer can do this with the proficiency of the human brain.

In our lifetime, we create thousands of conceptual models, all tied together in an intricate network of associations. Some mental conceptual models are very large, with many associations. Your mental model of your school and all that is in it is quite large now, but it will probably shrink after you graduate and other models become more important in your life. The brain, like all computers, is limited in its capacity.

To save time and space, the mind organizes our mental models hierarchically. Some models are larger and more important than are others. These models are selected and united by strong associations that form the rules and principles governing your actions and your worldviews.

Science introduces you to a lot of new and unfamiliar terms. Each term is associated with a conceptual model. Before science came along, the world seemed to humans to be a much more uniform place. Science by its very nature creates new concepts—new models of things that were not recognized before. The explosive growth in the number of new models over the past several centuries has led to increasing specialization, not just in the sciences, but elsewhere as well.

Today, no one person can know all of the models in all areas of science. The broad domains of science typical in high school (biology, chemistry, physics, and the Earth-space sciences) have become less meaningful in professional science, where the domains are blending, and then fracturing into specialized studies. Today, scientists in different specializations may not be familiar with one another's models.

Any given conceptual model is independent of the label that is put on it. Russian labels for many concepts are different from American ones. The concepts (in science at least) should be similar. Similar but not the same, of course. A conceptual model only exists—only has meaning—in our minds. We may create a model of it with images, words, numbers, and symbols, but that model means nothing without our interpretations. And those interpretations are invariably going to be somewhat different.

So in fact, there is no one model of any shared concept. There are as many models as there are people, usually with a certain agreement on major (core) characteristics. You and I would probably agree that a certain kind of animal is a cow, but our total conceptual model of cows might be considerably different. You and a nuclear physicist might agree on the most basic characteristics of an electron, but be worlds apart on the specifics.

Scientists are no different from other people. Scientific models may seem firmly established to you, because you see and learn one model. But the models you learn in science class are often highly simplified. They only approximate the greater complexities of the full models. The attributes of the full models may vary from scientist to scientist. You should not be surprised when you come upon different models of the same phenomenon in the literature you read. A model is always an approximation of its target.

For Discussion

1. Write the following sentence on paper: "A crippled old pine tree stood in the open field." Now visualize the scene. In your mind, find correspondences between each word and something in or about the scene.

2. Expand your mental model into a one-paragraph description, adding features. Compare your model to someone else's. Are they the same?

3. Select any model you see in the room you are in. What is the model representing about its target? What things in the target are missing in the model?

Student Reading # 4

The Models of Science: Description and Theory

There could be no fairer destiny for any physical theory than that it should point the way to a more comprehensive theory in which it lives on as a limiting case.

—Albert Einstein

Have you or one of your friends ever shrugged off an idea you didn't agree with by saying, "Oh, that's just a theory"? Most likely you meant that the idea was just a guess or didn't have much substance behind it. Most dictionaries agree with the way that you used the word. But as if often the case, the word *theory* has multiple meanings. In science, a theory, or, more properly for our purposes, a *theoretical model*, is just the opposite of a guess: it is a well-supported explanatory model.

A speculative, lightly supported model in science is called a hypothesis or hypothetical model. Some hypothetical models may grow and mature into theoretical ones, but more often, if they are verified, they enrich one or more existing theoretical models, as Einstein alludes to above.

The primary goal of most scientists is the creation of explanatory theoretical models. Of course, they create *descriptive models* of new phenomena or systems too. While description is important, it is usually only a first step. We can know something is there, but what does it mean? Explanation is considered the key to understanding any target system.

Natural scientists generally prefer models that include measurements and numbers for data, rather than just descriptions. It's easier, after all, to grasp the meaning of 30% of something than the meaning of "some" or "many." Science grew out of the application of mathematics to the physical world and mathematic models are still highly valued.

Occasionally, some of these mathematical models seem so enduring and stable that they are called *laws*. A law describes a mathematical relationship, but it does not explain it. A law may be an essential part of a theoretical model, but it is not the model itself.

Let's take as an example the law of gravity: a mathematical relationship describing the force of attraction between two objects. Without going into details or explanation, it is written as $F_g = G\ (m_1 m_2 / r^2)$. Scientists believe that this relationship holds true everywhere in the universe, which is why they call it a law. However, it explains little about gravity. A theory of gravity would explain why the law operates, what gravity is, and why it occurs. A viable theoretical model of gravity still eludes us.

Obviously, we could create our models simply by exploring, observing phenomena, and making logical suppositions about why things are the way they are. Isn't logic enough? Not really. That's the way learned people have reasoned throughout most of recorded history and you know where that has gotten us. The problem is that logic alone cannot compensate for errors in fact. Even the insane may reason with a strong logic, but their assumptions and their facts do not align with objective reality, so their subjective reality is not true.

Scientists reason from facts, but many nonscientific models, though logical, are based on intuition and unverified assumptions. If you assume that humans are the most important things in God's creation, then logic would dictate that our Earth sits at the center of the universe. This is what medieval scholars believed as they reasoned from analogy rather than fact (the Earth is to the observable universe as humans are to creation).

The conclusions we draw are called inferences. If you see your best friend walking toward you in the hallway with a sour look on her face, you might infer she's angry about something. At once, you will attempt to build an explanatory mental model so that you can act appropriately. Your model will just be speculative—a hypothetical model—until you get more facts. The facts may disprove your model, in which case you have to create a new one.

If my hypothesis turns out to be correct, I can add to my model of my friend—the mental model that allows me to understand her and her actions. I have a theoretical model of all my friends—in fact, for all of all people I know with whom I interact—and they have one of me. Those explanatory models allow me to predict how they will act in different circumstances. One friend may be stable and thoughtful—easy to predict—while another might be more volatile.

In science, a theoretical model is a model that has met the tests of *consistency* and *predictivity*. Scientific theories are largely *factual*, but they are not facts. A *fact* in science is an observation recorded as data. Scientific theories are inferred from facts, so they are factual.

Our test of a hypothetical model tells us how well it works—how well it explains what we observe. Its *truth* is how well it lines up with observed reality. And since we can never test all possible instances of a target, we can never be sure that our models will work under all circumstances. Scientific models can always be changed if they stop working, or if another model works better.

Theories are mental models. No one keeps a complete and written copy of all of the theories of science. There is no central repository of scientific models. The models you read in books are extremely simplified versions, as the authors understand them. Now, I should hasten to point out that just because everyone understands a given theoretical model differently does not mean there are no commonalities. Most of us agree on the basics of a story in a book, but we may have different opinions about the particulars of the story. The story evokes different emotions and images in each of us, but we all agree on the plotline. Just as no two readers understand a novel in exactly the same way, so no two scientists understand a theory alike. If you understand this, then you can understand why scientists might legitimately disagree on aspects of their models without rejecting the core principles of the model.

Scientists affirm their models rather than proving them beyond doubt. They may also falsify models by showing that they contradict observed facts. Even if it is highly unlikely, no one can be sure that an exception to a given model will not be found. All scientific models can be proven wrong.

For Discussion

1. Some scientists have said that laws such as thermodynamics and gravity are examples of models that have been completely proven. How can you know that exceptions do not occur somewhere in the universe? What assumptions are you making?

2. Based on this explanation, why do we not refer to the Law of Evolution rather than the Theory of Evolution?

3. What is meant by the statement: "In science then, any given theoretical model can be viewed as the sum of the theoretical models held by scientific experts in that field"? Think about it in relation to some nonscientific model you might hold in common with others in your school, like a model of "school spirit."

4. Why can scientists falsify models, but never prove them with absolute certainty.

Student Reading # 5

The Goals of Science

Man masters nature not by force but by understanding. This is why science has succeeded where magic failed: because it has looked for no spell to cast over nature.

—Jacob Bronowski

If you are fortunate, your school science experiences have included opportunities to design and conduct inquiries. *Inquiry* means literally "to look into." It is a crucial part of science along with exploration, discovery, experimentation, and research. Each of these terms means something different. In this reading, we will explore why do scientists do what they do. What are the goals of science?

If you've been taught that the goal of science is to explore and discover, you have part, but not all, of the answer. Scientists do explore and discover some of the time, but that is not all there is to science. Christopher Columbus, Vasco da Gama, Cortes, and Marco Polo were all explorers but they are not usually thought of as scientists. On the other hand, NASA's Mars Rover and Voyager programs and similar programs of the European Space Agency are considered science. What is the difference? It's not just the technology.

The difference is that scientists discover and explore to construct *descriptive and explanatory models* for their own sakes. These models, in their final form, appear as *research reports*. Contrary to what you may have been taught, science is not focused on invention, or finding solutions to solving practical problems. Those goals belong to technology, which is the subject of another paper. Trust me for now.

Okay. So what is my basis for claiming that the goal of scientists is to build models?

From the earlier papers in this series, you know that all meaningful knowledge exists as mental models. Whenever you learn something new, you must construct a mental model of it. This model is built by the processes within our brains using both the information we take in through our senses and from our existing mental models—memories.

We cannot meaningfully understand any of our experiences without referring to our existing mental models. Once we have constructed our mental model, then we can use it as a template to build an *expressed model* in whatever medium we are using: in words on our computer; on paper; in a verbal report; or in some other medium such as wood, clay, plastic, and so on.

This is true no matter what we are studying. Historians create models of the past; sociologists create models of social structures; artists create models to portray objects, ideas, and emotions. Scientists construct factual models to describe and explain physical reality. They do not have to have any purpose beyond that. Good scientists love to build good models, just as artists love to create in whatever medium they work in. Philosophers also create models of reality, but do not always restrict their efforts to factual models. Science originated as a form of *natural philosophy* that created and accepted only those explanatory models that could be tested and verified by facts.

Scientists, of course, do not have any outside authorities, like your teachers, to tell them whether their models are right or wrong. They rely instead on others in the *scientific community* to review their models and either accept or reject their models. While anyone may read and comment upon these models, other scientists with specific expertise in the subject being studied will carry the most weight in determining whether a given model is scientifically valid. These experts will examine the model to see whether the methods are sound, and whether the claims made in the model follow logically from the data.

Scientists usually build a model with some idea of where they want to end up; that is, they know the model they want to build before they start. Scientists do explore and they do discover, but once a discovery is made, they have to construct a meaningful model of it to present to others. Scientists don't tinker around in their labs until they happen upon a discovery. Discoveries that happen by chance (a process called *serendipity*) are by definition lucky accidents.

Very few scientists make discoveries that lead science in startlingly new directions. Most of them engage in what we might call *routine science*, focusing on the gaps and holes in the existing theoretical (explanatory) models to fill them. On rare occasions, their efforts lead to some sudden finding, insight, or understanding that an existing explanatory model cannot account for. They then must substantially modify the model or create a new model to explain the new facts. But the focus is on creating the model. The discovery comes before that.

Pulsars are rotating stars that emit beams of radiation in a pulsing manner, like celestial lighthouses. When the first one was detected in 1967, the scientific models then in place could not explain the extremely regular pulsations. Some scientists (and many proponents of extraterrestrial life) speculated that the signals might be coming from an alien civilization. Over the years that followed, though, scientists created a simpler new explanatory model that eliminated the need for alien civilizations to explain the phenomenon—a disappointment to searchers for extraterrestrial life.

Scientists build models through research and experimentation—two different but related activities. Not all research is experimental. *Research*, as the root of the word implies, is a systematic search. You may have done library research in English to find out about a particular topic. Research is more directed than exploration, in that it is focused on a particular goal.

Experiments include an element of deliberate testing; scientists conduct them to test their models. The test often involves a *prediction*: if my explanatory model is correct, then action X will lead to outcome Y. Experiments are a particular class of research, but not all research includes experimentation.

The ultimate goal of all science is the production of factual, research-based models of physical phenomena that withstand critical examination by scientific experts. Their research reports are expressed models including their actions and findings. It's not enough just to discover a thing: You also have to describe or explain it accurately. You have to create an argument for it. We'll talk more about that in another paper.

For Discussion

1. Does the explanation of science in this paper clarify the role of scientists for you? If not, what is confusing?

2. How are a scientist's expressed models different from those of a novelist? What is each trying to achieve? What assumptions do they make?

3. A scientist is studying the reactions of wolf spiders to sudden loud sounds. Follow the model-building process as the scientist completes her project.

Student Reading # 6

Science and Philosophy

I believe there is no philosophical high-road in science, with . . . signposts. No, we are in a jungle and find our way by trial and error, building our road behind us as we proceed.

—Max Born

Philosophy often gets a bad rap from scientific types, but to understand science and its relationship to society today, you have to understand the philosophy behind it. A *philosophy* is any system of logical principles that tells us how to behave and why we should behave that way. It underlies the complete composite mental model that is our worldview.

By understanding the philosophical tradition of most traditional cultures around the world, we can better understand some of the controversies about science in our society today. We all have a personal philosophy by which we live. It stems in part from our cultures. We cannot separate ourselves from that, and from our past, and pretend that we are so much different from people who lived five hundred years ago. Our lifestyles have changed, but many of our attitudes have not.

Culturally, European science evolved in the context of two competing philosophies: idealism and realism. The *idealist* philosophy is represented in the models of the classical Greek philosophers Socrates and Plato. Plato's idealism, sometimes called objective idealism, had a strong influence on the development of the Roman Catholic Church through the writings of scholars such as Thomas Aquinas.

The idealist model holds, in general, to the principles that (a) ideas are more important than physical evidence; (b) the mind exists independently from the body (as a soul, usually); (c) the mind can independently influence the physical world; and (d) a higher intelligence exists in a hidden world and is the source of ethics, morals, and ideals. Plato believed a world of ideals—of what he called forms—had a real existence and provided the templates our worldly reality is based on (the term "formal" refers to ideas of the hidden mind).

Extreme idealists may suggest that objective reality we experience does not really exist but is merely an illusion. In this view, called subjective idealism, only internal reality exists: we are creating the world that we are living in through our mind. For such serious idealists, the path to truth is intuition, not examination of the deceptive physical world.

In contrast to idealism, *objective realism*, a somewhat extreme model of realism, holds that: (a) the physical world is all that meaningfully exists; (b) the brain is the source of all ideas, ethics, morals, and thoughts; (c) no mind exists independently of the brain; and (d) the mind cannot act on the physical world alone. To the objective realist, only the physical world exists or is important, and the only path to truth is through research and experimentation. Other realist models are not so extreme.

Many people today maintain a personal philosophy that mixes elements of idealism and realism. Realism, with its focus on the physical world, is more compatible with modern scientific assumptions than idealism, and scientists in general are more realistic than idealistic.

While it is tempting to conclude that all scientists must be objective realists, such is not the case. Nor are scientists all coldly logical and atheistic (nonbelievers in God). Some scientists are atheistic—perhaps more than the general population—but many are not; however, scientists as a group do not tend to adhere to beliefs that contradict their models. Science as a whole takes no position on the existence of God. The thing called science is not a centralized political entity. By tradition and common agreement, though, scientists shun explanatory models that rely upon miracles, divine interventions, and prayer to explain events.

Although there is no single philosophical position one has to take to practice science, reputable scientific journals and conferences generally insist that the models they accept for presentation be built according to standard rules demanding factual evidence and physical explanations. Intuition is allowed only when one is speculating.

Early science was dominated by mathematicians, natural philosophers, physicians, and amateurs who had a keen interest in contributing to the "new science." At the time, idealism dominated much of society and the universities. Francis Bacon's work, *Novum Organum* (literally, "new instrument," referring to the methods Bacon championed) was published in 1620 and roughly marks the beginning of modern science.

Early modern science reflected the transition out of an age based on faith. Natural philosophers such as Newton, Boyle, and Priestly dabbled in alchemy, astrology, and the so-called "dark arts." The 17th century that marked the beginning of modern science was also marked by intense superstition and the persecution of witches.

Gradually, though, superstition and mysticism gave way to emphasis on reason and *empiricism*, a focus on physical experience as a source of knowledge. *Deism* removed God from any but the most occasional interventions in human affairs. Realist philosophies grew stronger, but they have not replaced idealism in the populist worldview of the 21st century. Worldwide, those with scientific realist philosophies are still in the distinct minority.

Science today rides upon an undercurrent of popular nonscientific and unscientific beliefs often more deeply held and more influential than contradictory scientific models. Since it is impossible to disprove the existence of hidden worlds, deities, and occult forces, and because these forces underlie many traditional practices and beliefs, they are extremely stable. In addition, science is far removed from being able to explain all things. Science deals with the physical world—not the spiritual one, if such a world exists. The mysteries of existence, consciousness, ultimate purpose, origins, and meaning continue to puzzle humankind and are unlikely to be answered anytime soon using scientific techniques.

For Discussion

1. Where do you come down on the idealist/realist continuum? What are five things you believe in that cannot be seen? How would you characterize yourself?

2. Investigate the meaning of empiricism. Compare it to objective realism as we define it here. Is there a difference?

3. Do a search on the web for "objective idealism," "17th century," or "objectivism," and write in your own words a short summary of three to five ideas you can add to flesh out those already stated in this reading (focus on science, if you can).

Student Reading # 7

The Clockwork Universe

Even if there is only one possible unified theory, it is just a set of rules and equations. What is it that breathes fire into the equations and makes a universe for them to describe? The usual approach of science of constructing a mathematical model cannot answer the questions of why there should be a universe for the model to describe. Why does the universe go to all the bother of existing?

—Stephen W. Hawking

Humans have long wondered about the origins of the universe. Prior to the 17th century, medieval Europe and much of the rest of the world held a mystical notion of a world guided directly by a god or the gods and fate. These worldview models questioned the reliability of our perceived knowledge of the objective world and the ability of humans to control their own destinies. Medieval peoples lived a haphazard existence, died young, seldom traveled, and felt themselves at the mercy of natural forces that had a will of their own. They relied heavily upon received knowledge—knowledge from God channeled through the church.

This began to change with the rediscovery of Greek, Roman, Byzantine, and Muslim writings in the Middle Ages. By the 17th century, the medieval system was breaking apart. Many people were living in cities and engaging in manufacture. Europeans were exploring the world, discovering new lands and peoples, and the riches of the New World. The printing press was making books available to well-off commoners; stable national governments were evolving; legal rights (at least for some people) were more respected; and complex machines were being invented. People felt they had more control over their lives.

With these changes came new ways of looking at the world. Machines in particular took on metaphorical meaning for people. They began to see the world less as an amorphous, willful beast and more as a mechanism—a machine—with one wheel driving another.

Isaac Newton is often credited with the model of a *clockwork universe*, though the idea was not new. Newton was highly admired and well respected in his own time for his genius, and so his ideas gained popularity among the evolving middle class of scholars and inventors. He suggested that God had created the world, set it in motion, and then withdrawn from human affairs except for the occasional intervention to keep things running properly. His model cast God in the metaphorical role of clock maker.

This model took root among many intellectuals and leaders of the 17th century, leading to a philosophy called *deism*. To the deists, who were especially strong in numbers in England and America, God existed but did not intervene regularly in human affairs. Tom Paine, Ben Franklin, Thomas Jefferson, and many other Enlightenment philosophers such as Locke and Hume adopted deism. Deists generally accepted the act of *divine creation*.

The rise of machines and manufacturing led to more mining and an increasing interest in geology. People began to take note of extensive indications that the Earth was much older than had previously been thought. By the 18th century, well before Charles Darwin, natural philosophers were concluding that the Earth was far too old to be explained by the model of creation in a Garden of Eden. They proposed an alternative model that included many of the elements of evolutionary theory, but they could not explain why evolution might happen.

By the 19th century, some natural philosophers were suggesting that the universe originated through some entirely natural process. Predictably, the religious community fought back against what they saw as a threat to reason. One argument against *natural origins* was the watchmaker analogy popularized by William Paley in 1802. This argument, based on the model of the clockwork universe, supposed that someone finding a watch on a path would naturally assume from the form and complexity of the watch that it had been made by a watchmaker. No rational person could believe that it had made itself. Paley thus championed Newton's model of a clockwork universe, arguing that the complexity of the universe required for a creator.

Paley's argument was purely analogical: God is to world as watchmaker is to watch. He could provide no specific evidence of divine creation, but did not see a need to do so. The *argument from design* seemed evident on its face. This is an excellent example of a *teleological argument* in which a preassumed goal or purpose shapes a subsequent explanation.

We have since become more aware that the universe at the atomic and molecular level has a capacity for self-organization that the parts of a watch do not possess. Paley did not know this, of course. The structure of the universe is consistent with this self-organizing capacity, although the models for how life began are still speculative. Current models of the origin of the universe do not address the involvement of a higher intelligence. It's likely that scientific explanations will only ever be able to go so far.

Notice I speak of "origins" and "evolution" as two different things. You may find it logical and plausible to believe in the divine creation of an evolving universe, as many people do. That does not make the model right, of course. What is right and what is logical are two different things, depending upon your assumptions.

Science seeks a natural origin, which is its right and purpose. Scientists cannot say that there is no purpose to the universe; it can only say that no purpose has been detected and verified. Therefore, we cannot include purpose in our current scientific models.

For Discussion

1. Scientific models are true if you accept scientific assumptions. What are the three most important of those assumptions about the world and how are they different from Paley's?

2. What are some other examples of teleological thinking that appear in our everyday lives?

3. Explain in your own words why, if it is not possible for scientists to disprove teleological arguments, they do not accept them into their models.

Student Reading # 8

Creativity in Science

The principle mark of genius is not perfection but originality, the opening of new frontiers.

—Arthur Koestler

You may have the impression that science is not a very creative activity. If you do, you probably misunderstand creativity. So let's look first at what creativity is, and then examine the notion that science is not a creative pursuit.

Simply put, *creativity* is the act of fashioning something that is (a) uniquely new, (b) useful, and (c) meaningful. All three of these elements must be present in a creative act. But we often confuse creativity with fantasy when they are not the same thing at all.

When we create *fantasies*, we take what we know about the world and rearrange the parts without worrying whether our final product corresponds to factual elements in the observed real world or not. Fantasy can be creative, but it can also be bland and unoriginal, or interesting but useless.

Scientists create mental and expressed models to represent targets they believe exist in the objective world. This is not easy to do. If it were easy, science would have evolved thousands of years ago. The world does not yield its secrets easily.

Each scientific model is a story. Scientists are its authors. The story is that of the nature of the physical universe. It is nonfiction, of course, but that doesn't mean it takes any less creative skill to build than fiction. In fact, it is harder because of the constraints.

Like an artist or architect, the scientist has to imagine his or her creation before he or she begins to build it. The mistake that many people make in thinking about science is that scientists just look and see what they can find. That they discover things and that's it. This is a very simplistic and inaccurate model of science. Let's look at a story and see whether we can't put together a more satisfactory explanatory model for science.

Once upon a time, long ago (and today in some parts of the world), people imagined that evil spirits were responsible for illness. This made logical sense to them: bad spirits were like bad people or the animals that caused them harm, except that they were invisible. Some might be the ghosts of their ancestors, come to punish them for some infraction. They had done something to deserve to become ill—of that much they were sure. Good and evil were balanced in the world, and bad deeds begat bad ends.

Then one day, some people called natural philosophers discovered tiny creatures under a glass: creatures that were too small to see with the naked eye. They were apparently very common and there were many different kinds.

Aha! Was it possible that these small creatures were what caused illness? It was definitely a creative insight to break from tradition and to create a model that linked the tiny creatures they saw under the microscope with people's illnesses. Some people laughed and ridiculed them, and others feared the retribution of angry spirits. They preferred to stick to traditional explanations.

But the natural philosophers and the physicians who accepted the new model pushed ahead. Their problem was to figure out what kind of model they could build that would prove their idea. They had to imagine what a model would look like that would convince people beyond doubt. They had choices. They could see whether sick people had the little creatures inside of them that healthy people didn't. They could infect other animals with the little creatures to see if they got sick. They could find a common source of the little creatures (such as drinking water) and see whether infected water made people sick.

Any one of these models might be convincing if it was well done. For each approach, the investigator would have to control a number of factors and collect certain kinds of data. He had to consider how convincing each model would be to others, and how practical it would be to construct. Each model would be an argument, so he would put the same creativity into its design that a debater would include in arguments she was preparing for a debate.

Once he selected a model, he would have to build it by carrying out the research or experiment, collecting the data, and interpreting the data according to his prior expectations. He would have to be very careful to follow his plan once he got started, but adjustments might be needed. As long as he could justify making the changes without biasing his model, he would be okay. This, of course, also required a degree of creative thinking because most changes required sacrificing something and one had to look ahead to see how a change might affect the product: the model.

Once he had his data, he would be fully engaged in creating the model—the research report—that he would use to convince others of the truth of what he had found; which is to say the alignment of his model with the facts. He would take care to show how his new model was better than the one in place. If he was fortunate, everything would work out and he would explain the source of a disease. He might then be able to go on to build a theory of the cause of diseases—at least some of them.

As this story illustrates, the scientist has be creative at every turn. This requires effort, will, and hard work: true creativity requires work and insight. The best artists and novelists can tell you just how hard it is to create new and genuinely meaningful products.

To be a good scientist, you must know the models that already exist, have a good command of a given situation, and apply a solid understanding of the tools and procedures of investigation and model building. You can't get by without long study and practice, but that is true of any meaningful creative activity.

For Discussion

1. Compare the mental model of creativity to that of fantasy and explain why it is important for understanding to distinguish the two.

2. Creativity usually implies a polished ability to fashion a useful and novel product in a particular medium (the stuff you work with). What is the medium of scientists?

3. Make a tabular model that compares and contrasts the creative products and processes of engineers, teachers, sculptors, and scientists.

Student Reading # 9

The Art of Creating Meaningful Models

All models are approximations. Essentially, all models are wrong, but some are useful. However, the approximate nature of the model must always be borne in mind....

—George Edward Pelham Box

"Aw, I have another lab report to write."

Does that sound familiar?

"What do you want to get out of the report?" I ask tactfully.

"An 'A,' if I can," you reply sarcastically.

"But," I reply tutorially, "What does the report really mean?"

You walk away in a huff, thinking that I just don't get it.

But I do get it. Honest. My question was intended to be helpful. After all, how we frame things makes a big difference in our attitudes toward them. Whether we feel challenged or burdened.

Reports are no fun to write unless you see in them an opportunity to learn: to show what you can do. That's the point of this paper. Some assignments don't give you much leeway about what to include. In others, you might have more freedom, more say in designing the report, which means you are more apt to behave like a real scientist.

What are you doing when you write a lab report? You are creating an expressed model that includes an explanation of why and what you did, and then a description or explanation of the phenomenon you studied. You are creating this model for others as an act of communication. You cannot escape communicating in life, and if you get good at it, you have a better chance for success, whatever you do in life. So you should practice and learn how to create the best models that you can—of all kinds.

I'm always surprised how many university students working toward advanced degrees in science don't fully recognize the real goal of their research. They think it is to discover something. But discovery is only part of the process. What good is a discovery if it doesn't become part of a larger model?

The purpose of any expressed model is to communicate clearly. Here is the scenario: I have a mental model of something I have observed that I want to convey to you. I can only do that through an expressed model, even if I just tell it to you. I must include in my expressed model what I think you need to understand my mental model. If I include much more, I will obscure my message.

Literary writers such as novelists learned many years ago to abandon ornate, flowery speech for plain, expressive, and powerful language. Unfortunately, many writers still include too much in their models. The first rule for good model building is to include no more than is needed to get the job done. Remember, models serve a purpose.

Of course, there's also the danger of including too little. You have to become like Mama Bear and get it just right. That's where skillful model building comes in.

Since most people respond better to pictorial and graphic images than to propositions, you should always include pictures, graphs, tables, drawings, and similar models when they can be useful. You will need prose (words) to explain your model, but include a mix of other models if they are appropriate. Whatever you do, you must make sure your model is clear. Scientists have to tailor their models—their papers and proposals—to the precise specifications of journals and grant agencies. Rule two: stay on message and follow the rules.

Building a good model is a project, and like all projects, you must plan it before you begin. You must visualize the product and know where you are going. You must not change course unless you have to. This plan is your map. Rule three is to begin and proceed with the end in mind.

Most professional research models are similar in the parts they contain. A generic scientific report would include

- *Problem*: what problem the model is addressing: what it will describe or explain.

- *Justification*: why the model is important and significant: how it will add value to existing models

- *Hypotheses*: any hypothetical models and the justifications for them (if models are being tested)

- *Methods*: a clear description of the methods for building the model

- *Results*: the data, graphs, charts, tables, etc., with summary statements

- *Analysis*: an analysis of your data and the inferences you can draw from it. If you are testing hypotheses, your explanation of how they are supported or not supported.

- *Conclusion*: How the model adds to existing models; why it is important and what models could be developed next.

If you look critically at these elements of a typical scientific model, you can see that they are not so different from the elements that a good debater might include in a presentation, or that a lawyer might include as she prepares the summary of a case for a jury. Clearly, your model is an *argument*.

All scientific reports are arguments. They are based upon fact and reason rather than upon clever twists of logic, but they are intended to be persuasive. The goal in science is to win the approval of one's peers, but to win honestly, according to the rules, based on the merits of your case. Bad models in science are eventually detected because they don't work. Others find out, and if you cheated, you lose their respect. That's the difference between science and politics. Rule four is to strive to create a strong but respectable argument in your model.

If you follow these guidelines, who knows? You might even get that "A."

For Discussion

1. Discuss the idea that scientists are creating arguments. How do you think they agree on a final model if there is a conflict?

2. If scientists all have the same facts, how can they produce conflicting models?

3. Even if you don't want to be a scientist, how could learning to write a good lab report benefit you?

Meaningful Data, Inferences, and Factual Models

It is a capital mistake to theorize before one has data.
—Edward Conan-Doyle,
The Adventures of Sherlock Holmes

Scientists create their models from facts and factual logical inferences. A *fact* is any event directly discernable by the senses that can be recorded as *data*. Any statement based on fact is *factual*. A factual *inference* is a conclusion we draw from facts. *Assumptions*, also important to good scientific models, are relationships or conditions we believe to be true, but that are not proven in this instance (*proof* referring to the results of tests).

Got it? Then let's practice. Look at each of the following three statements. Identify the fact, the inference, and the assumption.

1. There are cirrus clouds on the western horizon.

2. It is going to rain soon.

3. Cirrus clouds in the west mean rain.

If you followed the explanation above, you will have identified statement # 1 as the fact, # 2 as the inference, and # 3 as the assumption. Put the sentences together and you get the following statement: If we *assume* that cirrus clouds in the west mean rain, then the cirrus clouds on the horizon that I see (fact) means we will be getting rain soon (inference).

But, you may ask, what about the things that we can't see but that we know are there? Nobody has ever seen an electron. We only detect the effects of electrons. Aren't electrons facts?

Well, no, not really. The scientific model of the electron is a factual one: it is based on facts and logical inferences but it is not itself a fact. A fact is something that can be recorded as data. We cannot record an electron as data. We use data to infer the existence of electrons. (Just FYI: a single fact is properly recorded as *datum*. *Data* is a plural noun. Increasingly, though, we use the word data alone whether it's one fact, or many.)

How about the existence of the planet Mars? Is that a fact? Yes. It has been observed directly by many people. It is a fact that Mars exists.

Evolution is not a fact: It is a factual model constructed of many facts and inferences. We infer evolution from the evidence.

Gravity is a fact. The law of gravitation was constructed from facts, but the law itself is an inferred model. We cannot observe the law of gravitation. It has no substance. It is not a fact.

You can take this to extremes, of course, and claim everything we witness is inferred, since we only perceive our mind's' interpretation of facts rather than the facts themselves, but you probably don't have to go that far, unless you are a purist.

The data we include in our scientific models are either quantitative or qualitative. *Quantitative* data are numbers; *qualitative* data are descriptions, pictures, and any nonnumerical information.

Most natural scientists prefer to collect quantitative data because numerical measurements can be quickly and easily combined, averaged, and compared, in contrast with qualitative data. Numbers can be analyzed statistically; allowing us to estimate, for example, how likely it is that an observed effect is due to chance rather than the real effect of a treatment.

Numbers provide us with a sense of certainty, but they can be misleading if that is all one looks at. Numbers mean little if you don't really know the significance of your measurements, or if you are measuring the wrong thing. If you are measuring variables that are unstable, then the numbers are likely to be different when someone tries to replicate the model. Such a model is said to be *unreliable*. And statistics, while sounding impressive, may mean little. It sounds impressive to say that X happened 80% of the time, but if only five measurements were taken, the statistic is misleading.

Scientists usually construct qualitative models of a system before they can find and test meaningful components quantitatively. Knowing precedes explanation. If a target has many unstable qualities that change together as a system, teasing out a stable predictable relationship may be difficult or impossible. Scientists working with unstable systems must often content themselves with finding general predictive patterns in the data. They focus on estimating the *probabilities* of events, as meteorologists do to predict weather and climate change; as economists do in predicting trends; and as biologists do when predicting animal behavior.

Good data are the lifeblood of science. All scientific data have to be open to verification by repeated observation and testing. The following are NOT good scientific data sources:

- Unverified stories, no matter how logical or authoritative they may sound

- Unrepeatable or untestable observations

- Analogical and teleological arguments

- Uncontrolled or poorly controlled tests

- Uncontrolled or informal observations

- Testimony by untrained or unknown eyewitnesses

- Secondhand testimony

- Sources with a special interest in a specific outcome

• Witnesses under pressure or stress

• Self-reports of feelings or perceptions

These sources may provide clues that could lead to good models, but in and of themselves are of little value as the basis for scientific models.

For Discussion

1. Sit down with several classmates and independently describe any given object, such as a tree or some reasonably complex object in the classroom in no more than three sentences or three lines—but no numbers. Compare the results. Create a model of the object using these qualitative data. How did you create your model? What problems did you have?

2. Explain in your own words why it is important for understanding to distinguish facts, inferences, and assumptions. Make up a statement with all three.

Student Reading # 11

Sampling and Generalizing

Nature will not be Buddhist: she resents generalizing, and insults the philosopher in every moment with a million fresh particulars.

— Ralph Waldo Emerson

As you will recall from other readings, models are always simplifications of their targets and they are constructed for limited purposes. This principle holds true for all models that we create, whether we create them in our minds or with our hands. Every scientific model has one immediate target: the phenomenon the scientist studies directly. But that immediate target may be intended to represent a larger system of which it is a part.

To see how this works, let's consider the typical acid-base reaction lab you might carry out in your chemistry class. It contains simple, representative models of acid-base reactions constructed under uncomplicated conditions. In other words, it is an *exemplar*: a representative of a larger and more complex class of reactions. It does not represent acid-base reactions under all conditions.

Scientists usually study *samples* of a target that is too large to study in its entirety. That's what you do in your chemistry course. You could not study all acid-base reactions. You don't have the time or the resources to do that. Instead, you study a limited sample of them and use that sample to represent the larger class of acid-base reactions.

You are probably familiar with public opinion pollsters. They use samples of no more than a few thousand people to determine the opinions of whole states, or even of the nation. If they succeed, and they often do, it is because they construct a *representative sample* (an exemplary sample) that mirrors the target population in the most important ways. Ideally, the sample is the target population in miniature.

The trick to good sampling in any research is to match the important characteristics in the sample to the same characteristics of the target group. You can then create a model from the small sample that can be *generalized* to the larger target. To study the effects of differences in salt intake on the weights of elderly white males (EWMs), for example, you must select a sample of EWMs that is large enough to match the important physical characteristics of the population of EWMs in general, such as average starting weights and states of health, for example.

Only when you have such a match can you generalize the model from the sample to the target (all elderly white males) with any degree of confidence. But if you then generalize your model from EWMs to elderly white females, or to male in nonwhite groups, you run an increased risk of being wrong. And if you further generalize the findings to the entire population—young and old, male and female, all races—you would have a much greater risk of being wrong. It would be bad science, but it has been done in some medical studies.

Generalizing from a model based on a sample to a larger but similar population is common in scientific studies; and, of course, we are not just a talking about population models, but about all kinds of scientific models. We take samples of ocean water temperatures and generalize to the oceans as a whole. We have to do this. Studies of whole systems would be prohibitive in cost. But generalization always carries risks. A whole system may not behave the way a more limited sample does. In general, a larger sample size ensures better representation than a small one.

A second kind of generalization uses a model of one system to represent another, less accessible system. This is a risky venture because variables in the two target systems may be considerably different, so conclusions reached about one system may not be apt in the second system.

Medical and pharmaceutical scientists often use small animals such as guinea pigs, mice, and lab rats as subjects when testing new procedures or medicines. Ultimately, these tests also have to be conducted on humans. Because of differences in the body chemistries of—say—a lab rat and a human, many potentially exciting medical breakthroughs in areas such as cancer treatment have folded when successful models from animal studies proved to be invalid in human trials.

Scientists are usually careful to create models that accurately represent their immediate targets, and cautious when they generalize to other targets—especially targets that may be different in significant ways from the systems they use for their models. Models are always simplifications of their targets; the greater the number of differences between two target systems, the greater the likelihood of error in using one system to represent the other.

Still, as long as we are mindful of the danger of overgeneralizing, we can often use models from a studied system to suggest possible features of another, unstudied system. Knowing life exists on planet Earth, we might reason that life would exist on another planet with the same physical characteristics. This kind of generalization allows us to use Earth as a standard model to focus our research.

The features of a single-celled creature such as an amoeba allow it to live independently. It is a complete cell. If you create an expressed model of the amoeba, you can generalize the model to understand cells in other conditions, but with limits. The further you get away from the free-living conditions of the amoeba, the less representative the amoeba becomes. Thus, the model of the amoeba would be relevant to understanding other cells in a hierarchy:

- Other protozoans in general would presumably be most like the amoeba

- Cells of algae, except they are photosynthetic and do not hunt food

- Cells in multicellular animals, except they are usually specialized and not free-living.

• Cells of multicellular plants, except they are sometimes photosynthetic, sometimes not, and are specialized and not free-living.

At each level, the amoeba model *fits* less well with the cells found there. The hierarchy is rough, though. Some human white blood cells are more like amoeba that some protozoans swimming next to it in the same pond. We cannot know for sure how representative a model is of a particular target without studying the relationship between them.

The point is that you must be cautious about generalizing while at the same time you must recognize the need to generalize. Generalizations beyond the original target of a model need to be tested. And the farther away from the original target we go, the more cautious we must be about accepting a generalization at face value.

For Discussion

1. Think about a lab you may have done recently and create a set of generalized targets for it as we have done with the amoeba in the reading. For each target, identify possible differences between the model and target that would limit the model's usefulness for describing the target.

2. Scientists often include speculative generalizations in their models to point toward new and possibly fruitful research. In the model you just created, can you see any new questions for research?

Student Reading # 12

Inductive and Deductive Models

The very hope of experimental philosophy, its expectation of constructing the sciences into a true philosophy of nature, is based on induction. . . .

—Chauncey Wright (1871)

To understand the historical development and practice of science, you must understand inductive and deductive reasoning. Both are essential to science, but, as Wright says in the quote above, induction is crucial to science. The differences between these two forms of reasoning are not difficult to understand.

Simply put, *induction* requires you to use facts to construct a general explanation. With *deduction*, you use a general explanation or principle to interpret the facts. Simple, right? Well, maybe not. Let's go on.

Wright's idea that a "true philosophy of nature" (e.g., science) could come about only through induction was a reaction to a tradition of medieval scholarship called *scholasticism* that dominated European thought for centuries. In medieval universities, scholars learned the models of the past—mostly the works of philosophers such as Aristotle and physicians such as Galen—and created arguments to justify them. In other words, they sought to make the facts fit a pre-existing model. This created increasing conflict as more and more facts became known, and eventually led to the intellectual revolt that we call the scientific revolution—part of a general movement called the Enlightenment.

The natural philosophers (a.k.a., early scientists) were intellectual revolutionaries. They wanted to begin with the facts to create models that fit the facts, which today seems to most people like such an obvious and good approach that we might be astounded that anyone would think otherwise. I hasten to point out, though, that many people today cling to models that contradict the facts, so perhaps we are not as advanced in our thinking as we would like to believe.

In 1620, Sir Francis Bacon, a retired Lord Chancellor of England, published an important book called *Novum Organum* (literally, "new method") in which he championed the use of induction for investigating the physical world.[1] The method became known as the Baconian method, although it was around well before his time.

Most modern scientific models begin with induction. Investigators assemble a model of a system by studying the facts. From these facts, they make inferences. We do this all the time in our daily lives. The key, though, is to assemble the facts and create the model without interpreting the model according to some organizing principle that would bias our interpretation of them. Bacon's idea was to let the facts speak for themselves.

---

[1] Another Bacon, Roger Bacon, championed empirical (hands-on) methods of research in the 13th century. There's no evidence he was related to Francis Bacon.

In reality, there is always an organizing principle. Facts can't speak for themselves. A better interpretation of induction is that the facts are not forced to fit a model that is already in place. The facts lead to the model, rather than vice versa. In police work, this would be interpreted as finding a suspect by following the facts, rather than making the facts fit a preconceived suspect.

Once the facts have led to a model, then the model is tested using deduction. Deductive methods are "if . . . then" propositions that begin with the explanation and lead to more facts. To go back to the police work analogy, once a suspect has been identified inductively, then the "theory of the crime" must be tested by finding facts that support it. All facts may lead to the inference that A committed the crime, but if A has an airtight alibi, the model folds. Deductive testing is essential to affirm a model: "If A committed the crime, then X, Y and Z will be true."

In science, we create *hypothetical models* from our inductions, and we test these models to see whether the facts will support them. We say, "If H is true, then X, Y, and Z will be observed." Our hypothetical models may be little more than a hunch based on a few facts: just as a theory of the crime may be a poorly supported hunch. Notice that deduction is essentially the same thing as testing predictions.

Descriptive models are generally created through induction. They describe what exists, and can be tested by independent replication. It is the inferences—the explanations for these models—that we test deductively. A purely descriptive model is tested through replication.

By making and testing predictions, we test the model. The Mars Rover sends back photographs of the surface of that planet and other data measuring temperature, light intensity, and so forth, we can create a descriptive model of its surface. We may infer a model from clues as to how the features were formed. We might create a hypothetical model that says certain features as formed by running water. We might then predict that water would be found below the surface of certain features if our speculative hypothetical model is valid. We now can search for water to see whether our prediction is correct. If it is, it validates and reinforces our model, but does not prove it conclusively. If enough predictions prove correct, though, we end up accepting the hypothetical model and we add it to our theoretical model of Mars.

This is a simplified model of how science works by using a "scientific method" called the *hypothetico-deductive method*. In the real world, no one method is used by all scientists all of the time. Models of events that have already occurred once (such as the origin of our universe) may not be testable by controlled experiments, but it may still be accepted based on the weight of supporting factual evidence and logical inference. Testing lies at the heart of science. Unless they are tested, inferential models are speculative, no matter how established they may seem to the nonscientist.

For Discussion

1. Give an example of how you might create inductive and deductive models in your daily life; say, in your relationships with friends of family.

2. Look up the Baconian method on the internet and summarize what it is in three to five sentences.

3. Imagine you are a good mechanic and your car won't start. Explain what you would do to apply the hypothetico-deductive method to arrive at a cause.

Student Reading # 13

Science and Technology

Central to the great scientific revolution of the seventeenth century was the interpenetration between the technical and the scientific. For better or for worse, that interpenetration marked all of western civilization and took on a form . . . it did not have in the ancient and medieval worlds.

—Paolo Rossi, *The Birth of Modern Science*

The Greek term *tekhnologia*—the root of our word, *technology*—referred to the systematic study of art, craft, or technique. Although technology may seem to refer to machines and mechanics, it actually refers to the study of *techniques*, which are specialized methods and procedures. By this definition, some philosophers argue that science is a form of technology. After all, science is distinguished by its techniques. However, other philosophers argue that science should be distinguished from technology because of its goals. They say that technology is the domain of practical problem solving and invention, while *science*, in contrast, is the hunt for descriptive and explanatory knowledge, a.k.a., mental and expressed models. Science in this view is not an art or craft in the sense that it does not produce a physical artifact.

Sometimes, science is broken into two subcategories: applied science and so-called "pure" science. *Applied science* blends the characteristics and goals of technology and science, while *pure science* is science as we defined it in the previous paragraph. As this brief discussion makes clear, the relationship between science and technology is the subject of debate. We will not settle it in this book, but we will discuss later why a distinction may be important.

Technology predates modern science by many thousands of years. The technology of the prescientific past was practical, focused on the arts and crafts needed to meet basic human needs such as clothing, shelter, food-getting, and security. For the most part, these early technologies developed through trial and error rather than any understanding of the phenomenon being exploited.[2] Explanations for natural phenomena were couched in mystical terms: Elements had souls and steam rose because it had a need to fulfill. *Animism*—the idea that all things have souls, purpose, and unique behaviors—was a dominant model.

As important as technology was to the ancient civilizations, by Plato's time (around 428–328 BCE) many philosophers and mathematicians looked down upon the mechanical trades, preferring to seek truth in the world of ideas (a philosophy called *idealism*). Mathematics and the arts were valued, and applications

---

2   Thomas Edison used the same kind of trial and error to discover a working material to use as a filament in his light bulbs since he had no explanatory models that could guide him in his search. The theoretical understanding of electricity and conductors was still in its infancy.

of mathematics, but the physical world was generally regarded as unpredictable and illusory. Most philosophers tried to explain the world without testing their inferences.

Complex machines are known from the ancient world, but most of these appear to be novelties. The world of abstract ideas was held in higher regard that the world of practical invention and this appears to have been true across many ancient civilizations. Aristotle apparently went beyond others in his studies of nature, but mathematics, physics, and astronomy dominated the ancient *protosciences*[3] because the ancients believed that mystical truth and power lay in numbers, shapes, numerical relationships, and astrology. The relatively modern notion that 13 is unlucky or that 666 is associated with the devil are instances of *numerology*. Chemistry was largely studied in the context of alchemy and biology in the context of medicine.

Trends toward more scientific thinking came about slowly, championed by occasional individuals but largely ignored by the establishment, which was vested in tradition. No one wanted to offend the deities or vengeful spirits by questioning their very existence, and the only way to explain how things worked was through analogies with the purposeful world they could see: the human world. Even technology took a long time to develop relative to the span of human history, but the tipping point was finally reached beginning in the 17th century in northern Europe.

Many circumstances came together to set the stage for the development of the "new science": the development of the printing press, more orderly government, greater intellectual freedom, the rise of capitalism, the development of a middle class, industrialization, a booming interest in machines and invention, the wealth of the new world, and exploding knowledge of the world beyond the confines of Europe. Intellectuals rebelled against the explanations and restrictions of the past as a spirit of innovation took root.

A growing respect for innovation fueled the birth and evolution of science. It was no longer shameful to get your hands dirty building machines; no longer foolish (or dangerous) to probe into the mysteries of nature. Technology was essential to the new science, but for a long time, science was not important for technology. Even well into the 19th century, much science was carried out by amateur natural philosophers who were in their real lives physicians, clergymen, businessmen, manufacturers, landed gentry, and the like. Darwin was a wealthy gentleman, while Gregor Mendel was a monk and high school teacher. The modern use of the term *scientist* came about only in the early 19th century. Much of the world was still unknown to Europeans including large swaths of interior Africa. Science mixed with the supernatural in popular literature.

Gradually, science began to change technology. Engineers began to study scientific models, using them to create new products. The science of medicine

---

[3] A term used to describe activities that ultimately led to science but were not scientific in the modern sense.

developed in earnest. Old analogical explanations lost their power. Science and technology came together and joined as one. But there are reasons today for wanting to keep science and technology apart. Science is concerned with building descriptive and explanatory models, without regard for how the models might be applied. Scientific models are ideas. Modern principles of free speech are based upon the assumption that the free exchange of ideas is in itself not harmful. Acting on ideas is a different story.

Scientific models are ideas; technological products and inventions are actions. Modern efforts to maintain the distinction between science and technology are thus based in part on the desire to maintain the freedom of expression that is necessary for science to thrive.

For Discussion

1. How has science changed our ideas about ourselves and our world in comparison to the medieval worldview? How about technology? Can you identify specific ideas that have had a real impact on our culture?

2. What is the relationship of mathematics to science and technology? Mathematics is generally not considered a science in the modern sense. Can you see any reason why it is not?

3. Medical science is distinguished from medicine in this paper. Why would medicine be considered a technology rather than a science?

Science and Myth

Alice laughed. "There's no use trying," she said; "one can't believe impossible things."

"I daresay you haven't had much practice," said the Queen. "When I was your age, I always did it for half-an-hour a day. Why, sometimes I've believed as many as six impossible things before breakfast."

—Lewis Carroll, *Through the Looking Glass*

Myths and *legends* are fictitious stories that have symbolic meaning to the listener or reader. They pervade all cultures, and some of them are based upon true stories or events: others are obvious fantasies. Simple errors or misunderstandings are not myths, although you may find some writers using the term *myth* this way.

In ages past—and still today—our myths and legends were our primary means of understanding the world and our place in it. They were our models of existence. History as we know and study it today did not exist for most people; instead, they lived in stories rich in symbolism and emotional appeal. No one knows whether the original storytellers believed their own stories. Some may well have convinced themselves that they were receiving the wisdom of the gods or the ancestors. Others may have been more deliberate and cynical. But regardless of the source, listeners often believed their stories to be literally true.

And the magical beliefs of most cultures in the past were compatible with mystical and supernatural models many people find incredible today. Yet myths and legends persist today in even the most scientific of societies. They undergird commonly accepted explanations, rituals, rules, and practices, and they serve to keep tribes, clans, societies, and nations together when danger threatens. The causes of most wars are in part mythical, and soldiers often fight for mythical ideals. Myths also underlie many of our social interactions. In fact, the amount of myth that underlies the cultural beliefs of most societies—even "scientific" societies—is astounding.

Myths are different from nonmythical and scientific models. Mythical models appeal to the emotions and often attain resonance (harmony) with important, commonly held mental models in a culture. The warrior myth, for example, underlies the attraction of some people to football; while the princess myth underlies the willingness of people to lavish money they might not have on an expensive wedding. Myths and legends serve as models that impart lessons to us, inspire us, tell us how to behave, and how to live our lives through analogies. Once they become part of our mental models, they are often difficult to get rid of because we become so emotionally invested in them. This stems from their symbolic function. Football players are not warriors in the true sense, nor are most brides princesses.

Culturally, myths serve an adaptive purpose. Mythical stories define good and evil. They represent something higher and better: they are models to which we can aspire. They provide us with life's lessons in symbolic form, often peopled by exaggerated heroes and villains.

While science has given humanity many gifts, it has also destroyed—or at least has attacked—many myths. Darwin's impact was not just on science, but upon all humanity. By explaining natural selection, Darwin made the divine creation myth much less tenable, just as the advent of computers has made the idea of a human mind separate from the body more difficult to defend.

As might be expected, modern opposition to science come primarily from groups that have a vested interest in maintaining traditional or more recently created myths. Myths, while often set in the past, may be created at any time. The *entertainment legends* of the *Star Wars* movie series and the *Lord of the Rings* trilogy, while set in another time and place, are of more modern origin.[4] Most *religious myths* are set in the distant past, though some have originated relatively recently. *Ad myths*—myths created to sell products—are also recent, as are *urban myths.*

Mythical models are by definition nonscientific (because they are fictitious, and science does not deliberately include fiction in its models) or unscientific (contradicted by scientific models). Not all stories are mythical. Some may merely entertain us without unusual symbolic value. A person with a scientific worldview values truth (meaning the harmony of her mental models with observe fact) and so tries to distinguish mythical from real models. She may value myths for their entertainment value and even appreciate the lessons they teach, without accepting them as real. Many people appreciate the symbolic value of religious myth without accepting the stories as literal truth. A scientifically literate person may appreciate the rich symbolism of myth in a film, just as he may appreciate speculations on the supernatural found in good science fantasy without making it part of his real-world model.

How do we tell mythical models from reality models? They are not always obvious, but in general, myths tend to

- be rooted in the past, often in traditions dating back centuries;

- have no specific authorship or source;

- involve elements of magic or the supernatural;

- have symbolic meaning that ordinary stories do not have;

- feel too perfect or neatly moralistic;

---

[4] An author may not deliberately give a story symbolic value. The symbolism may evolve later. Elvis Presley's life has taken on a certain mythical status for some people but not for others, for example.

- feel exaggerated or "bigger than life";

- support a cause or special interest;

- include unnatural, unlikely, or even impossible events;

- resonate with, and reinforce, established beliefs;

- contain strong elements of analogy with ideal models; and

- lack documented supporting evidence.

Most of us can recognize myths by exercising healthy skepticism. Mythical models normally exhibit more than one of the characteristics above. Commercial myths support the sale of a product. Hero myths provide us with icons to emulate. Founder myths glorify the founding of a country or cause. Social myths support a particular social structure. Religious myths support a religious worldview. Urban myths symbolize our fears of some feature of urban society. Personal myths support our personal views and actions. You get the picture. The essential feature of all myths is that they have no supporting facts and therefore can never be scientific.

For Discussion

1. Myths are pervasive in areas of our lives such as gender relationships, religion, race relations, child raising, partnerships, friendship, and so forth. Try to identify five beliefs you hold in any area that could be myths or that you know are mythical.

2. Science busts myths, but has its own mythical component. Do a web search using such key words as "Galileo myths" or Newton myths" or "Science myths" and see what myths you can find for discussion. Keep in mind that myths are symbolic stories—you may find the term *myth* being applied erroneously to simple misunderstandings. They are not true myths.

Student Reading # 15

Skepticism and the Scientific Worldview

The same principles which at first view lead to skepticism, pursued to a certain point, bring men back to common sense.

—George Berkley

Nobody likes a skeptic, especially the dishonest sales personnel trying to close a sale. This is not to put down anyone in sales; but the good consumer must be a skeptic. And you are always going to be a consumer. As you learn from others, you are consuming their ideas and opinions. It doesn't hurt to maintain a healthy skepticism towards their models. *Skepticism* refers to the practice of critical questioning. It is not the same thing as *cynicism*, which also doubts human motives. The skeptic does not have to be a cynic as well.

We live in a world filled with misinformation and disinformation (false information or, more bluntly, lies and deceptions). Most people view science positively.

The scientific worldview values fact-based models over romantic *intuitive* models: models based on feelings. We may find the latter more satisfying in some cases, but following intuition over fact is more likely than not to lead us to make wrong decisions. There is nothing inherently wrong with intuition: it is a wellspring of creative thought and insight. By definition, intuition is independent of any conscious reasoning process.

It's a popular myth that scientists are coldly logical, free of emotions in their decision-making. Scientists tend to be more logical than the average person is, but they do not lack emotions. The term *emotion* means, "to set in motion." Without emotions, we would not be motivated. We experience emotions as feelings, and scientists are just as feeling as anyone else. What scientists try to avoid is *bias*: slanting a model in a favored direction. They may expect a model to turn out in a particular way, and they may be disappointed if it does not. They have feelings about the models they construct, just as an artist has feelings toward the works of art he or she creates. They take pride in them.

Even skeptics might be less prone to question models they feel good about. If a particular model feels good and works for us, we tend to accept it less critically. But the true skeptic knows that it is important to question all models equally, especially those that might be misleading because of their emotional appeal. In the scientific worldview, the facts rule. All else is speculation. Thomas Huxley famously stated that, "science . . . commits suicide when it adopts a creed" (Huxley 1885), but factual creeds are necessary and desirable in science.

We should not blindly trust science and scientific reports, any more than we should blindly trust any human endeavor or set of mental models. This is in part because most of the scientific models we receive through the popular media are selected and interpreted by nonscientists in the role of *informal experts.* Often they have a vested interest in popularizing certain stories, and they tend toward sensationalism to attract viewers or readers. Because they are interested most of all in attracting an audience, they may report stories prematurely, or without checking facts, or exercising normal journalistic skepticism. In fact, reporters may not have the expertise to ask the right questions.

These kinds of informal experts may have impressive credentials and sound like experts—but they tend to reply on their own intuition to determine what is true. An informal expert is someone whom the public perceives as expert, but who lacks the extensive background of research and study typical of *formal experts* in a field. Formal experts, in turn, often lack the communication skills (or interest in popular communication) of the informal experts. And formal experts in one field may be erroneously perceived as an expert in another different field, by virtue of being a scientist or holding an advanced degree in some related field. So we must be skeptical—if not cynical—about the source of our news and the credentials of those who present themselves as experts.

Another reason for being skeptical about scientific models is the nature of modern science itself. Few people question the success of science in explaining the physical world, but its very success has turned it into a profession driven by intense competition for financial support and other resources. Modern science has become ever more expensive as its targets become more remote and technically difficult to model. Expensive equipment, high operating costs, and the need for specialized services are the rule in some areas. This cost results in strong pressures for positive results. No one wants to pay two billion dollars for a project that fails to create a viable model. In addition, many universities demand that their faculty get grants and engage in successful research or lose their jobs, and the demand for success is always present in private institutions. Scientists in these situations may become desperate enough to engage in professional misconduct by making up data or biasing their models in other ways.

Private enterprises that may pay scientists to conduct research with the expectation that they will produce models favorable to the organization's needs or products. This is also true in some government agencies, where research models that contradict government policies may be quashed.

Most science is conducted honestly, by competent scientists who respect scientific ethics and adhere to high standards of scientific conduct. But given studies that show such misconduct occurs, how can you, as a skeptic with a scientific world-view, tell good models from bad? You probably can't with any certainty, but you should be cautious about immediately accepting scientific models that

- sound too good to be true (they probably are);

- are funded by a corporation or special interest group and support the interest or product of the funder—you can often find out a group's bias on the internet;

- are results of animal studies—such models often do not transfer to humans;

- are the results of only one test or a small group of subjects;

- are based on an uncontrolled observation study (because uncontrolled variables could account for the effect);

- have not been reviewed by scientific experts in the field;

- are based upon a small effect (it could disappear in further tests);

- make extraordinary claims (extraordinary claims require extraordinary evidence);

- contradict established scientific models (most such models prove wrong);

- affirm cherished cultural or social beliefs (possibility of bias); or

- are based upon subjective evidence such as people's feelings or self-reports (notoriously unreliable source).

Models with these characteristics could be affirmed with further work. A scientific thinker will not immediately reject any model that is not obviously unscientific. But she will be cautious about accepting the claims without further confirmation.

For Discussion

1. What is your worldview? How much of a tendency do you have to question what you are told?

2. Do you think intuition is the best guide to some actions? If so, what actions? Why would you favor intuition over fact?

3. Find several examples of scientific misconduct (search for "scientific misconduct cases") on the internet and examine them closely. Why did the misconduct occur in each case? What harm, if any, resulted?

APPENDIX 2

Recommended Resources

Many websites and books deal with topics covered in the book. For the most part, internet resources can be had by typing in key words and conducting a search. Type in "mental models" and you will have many sites from which to choose. I find *Wikipedia* does a good job of presenting most science-related topics. Often you can find websites taking positions on issues at odds with science. These resources are great targets to subject to critical analyses using MBST protocol.

The websites listed below are subject to change.

For case studies related to science, *The National Center for Case Study Teaching* at the University of Buffalo maintains a database of cases studies that you can integrate into your curriculum in every field. Studying these cases as models adds an extra dimension to them. Go to *http://ublib.buffalo.edu/libraries/projects/cases/ubcase.htm*

Snopes.com is a good place to go to find and to check out rumors and urban myths related to science, although some of these "myths" are just erroneous beliefs with no mythical status, as we explained in Chapter 7. Go to *www.snopes.com/science/science.asp*

SourceWatch is an excellent resource for teachers who are seeking information on contemporary rumors and issues. A service of the nonprofit, nonpartisan Center for Media and Democracy, SourceWatch "profiles the activities of front groups, PR spinners, industry-friendly experts, industry-funded organizations, and think tanks trying to manipulate public opinion on behalf of corporations or government." Not all have to do with science, of course, but many do. Go to *www.sourcewatch.org*

To identify hoaxes and frauds in science to discuss in class, check out these sites:

- Urban Myths: *www.urbanmyths.com*

- Scambusters: *www.scambusters.org*

- Museum of Hoaxes: *www.museumofhoaxes.com/hoax/archive/display/category/scientific_hoaxes* or *www.museumofhoaxes.com/hoax/archive/display/category/scientific_fraud*

- New Scientist: *www.newscientist.com/article/dn15012-seven-of-the-greatest-scientific-hoaxes.html*

- Listvers: *http://listverse.com/2008/04/09/top-10-scientific-frauds-and-hoaxes*

- Neatorama: *www.neatorama.com/2006/09/19/10scientific-frauds-that-rocked-the-world*

- American Thinker (When Scientific Fraud Kills Millions): *www.americanthinker.com /blog/2010/01/when_scientific_fraud_kills_mi.html*

Books of Interest

The following can be useful to you as a source of cases for your students to study and analyze, and for you as you seek to expand your expertise in the history and nature of science and in modeling as an educational approach.

Fraud and Misconduct

Gardner, M. 1957. *Fads and fallacies in the name of science.* Mineola, NY: Dover Publications.

Gardner, M. 1989. *Science: Good, bad, and bogus.* Amherst, NY: Prometheus Books.

Goldacre, B. 2010. *Bad science: Quacks, hacks, and big pharma flacks.* London, UK: Faber & Faber, Ltd.

Goodstein, D. 2010. *On fact and fraud: Cautionary tales from the front lines of science.* Princeton, NJ: Princeton University Press

Huber, P., and P. W. Huber. 1991. *Galileo's revenge: Junk science in the courtroom.* New York: Basic Books.

Park, R. L. 2000. *Voodoo science: The road from foolishness to fraud.* New York: Oxford University Press.

Seethaler, S. 2009. *Lies, damned lies, and science: How to sort through the noise around global warming, the latest health claims, and other scientific controversies.* Upper Saddle River, NJ: FT Press Science.

History of Science

Bryson, B. 2010. *Seeing further: The story of science, discovery, and the genius of the royal society.* London, UK: Harper Press.

Conner, C. D. 2005. *A people's history of science: Miners, midwives, and low mechanicks.* New York: Nation Books.

Grant, E. 1998. *The foundations of modern science in the middle ages: Their religious, institutional and intellectual contexts.* Cambridge, UK: Cambridge University Press.

Gribbin, J. 2003. *Science: A history: 1534–2001.* New York: Penguin Press.

Lindberg, D. C. 2008. *The beginnings of western science: The European scientific tradition in philosophical, religious, and institutional context, prehistory to A.D. 1450.* Chicago: The University of Chicago Press.

McClellan, J. E., and H. Dorn. 1999. *Science and technology in world history: An introduction.* Baltimore: Johns Hopkins University Press.

Science, Religion, Paranormal, and the Supernatural

Ferngren, G. B., ed. 2002. *Science and religion.* Baltimore, MD: Johns Hopkins University Press.

Goran, M. 1979. *Fact, fraud, and fantasy. The occult and pseudosciences.* Cranbury, NJ: A. S. Barnes.

Models and Human Thinking

Clement, J. J., and M. A. Rea-Ramirez., eds. 2008. *Model based learning and instruction in science.* New York, NY: Springer.

Gilbert, J. K., and C. Boulter., eds. 2000. *Developing models in science education.* Norwell, MA: Kluwer Academic Publishers.

Gilbert, J. K., ed. 2005. *Visualization in science education.* Dordrecht, the Netherlands: Springer.

Johnson-Laird, P. N. 2008. *How we reason.* Oxford, UK: Oxford University Press.

Harris, S. 2010. *The moral landscape: How science can determine human values.* New York: Free Press.

INDEX